Synthesis Lectures on Ocean Systems Engineering

Series Editor

Nikolas Xiros, University of New Orleans, New Orleans, LA, USA

The series publishes short books on state-of-the-art research and applications in related and interdependent areas of design, construction, maintenance and operation of marine vessels and structures as well as ocean and oceanic engineering.

Fidaa Karkori

Ship Vibration 1

Vibration Analysis Methods for Vessels

 Springer

Fidaa Karkori
Southampton, UK

ISSN 2692-4420   ISSN 2692-4471 (electronic)
Synthesis Lectures on Ocean Systems Engineering
ISBN 978-3-031-75071-7   ISBN 978-3-031-75072-4 (eBook)
https://doi.org/10.1007/978-3-031-75072-4

This Springer imprint is published by the registered company Springer Nature Switzerland AG
The registered company address is: Gewerbestrasse 11, 6330 Cham, Switzerland

If disposing of this product, please recycle the paper.

Preface

It is universally recognised and accepted that overall ship vibration is an important measure to ensure the habitability, safety and functionality of marine vessels. This book examines the causes and effects of ship vibration to provide users with specific guidance on the design, analysis, measurement procedures and criteria in order to achieve the goal of limiting the ship vibration to an acceptable level.

This Book 1 considers the design and construction of the hull, superstructure, and deckhouse of a steel vessel based on all applicable requirements of established classification society (Class) rules. Specifically, for container carriers over 130 metres in length, the Class rules require the consideration of vibratory responses of hull structures, as applicable. For LNG carriers, the Class rules require special attention to the possible collapse of the tank membrane due to hull vibration. For ship-type floating production units with spread mooring, the rules relating to the building and classing of facilities for offshore installations require the flare tower/boom and hull natural frequencies to be separated to avoid resonance or near resonance.

This book should be read in conjunction with Book 2, Propulsion Shaft Alignment, which considers the effect of propulsion shaft vibrations.

This book provides practical guidelines on the concept design to assist ship designers to avoid excessive shipboard vibration at an early design stage. These guidelines also assist with the finite element analysis (FEA)-based vibration analysis procedure to calculate the vibration response and evaluate the design at the detail design stage. The analysis procedure represents the current analysis practice used by Class.

These guidelines also offer guidelines on the vibration measurement procedure at sea trials and the acceptance criteria on vibration limits based on international maritime industry standards.

Southampton, UK   Fidaa Karkori

The original version of the book has been revised. A correction to this book can be found at https:// doi.org/10.1007/978-3-031-75072-4_10

Contents

Abbreviations and Acronyms

A/D	Analogue to Digital
AC	Alternating Current
BEM	Boundary Element Method
C_F	Conversion Factor
CG	Centre of Gravity
COMF	Comfort (Class notation)
COMF+	Comfort Plus (Class notation)
cpm	Cycles per minute
DC	Direct Current
FE	Finite Element
FEA	Finite Element Analysis
FFT	Fast Fourier Transform
FPU	Floating Production Units
HAB	Habitation (Class notation)
HAB+	Habitation Plus (Class notation)
HP	Horsepower
Hz	Hertz
ISO	International Standards Organisation
kN	Kilo Newton
kPa	Kilo Pascal
kW	Kilowatt
m/s	Metre(s) per second
mm	Millimetres
MRA	Maximum Repetitive Amplitude
MSDV	Motion Sickness Dose Value
PRU	Power-Related Unbalance
Rpm	Revolutions per minute
SNAME	Society of Naval Architects and Marine Engineers (US)

TBC   Top Bottom Centre
TDC   Top Dead Centre

List of Figures

List of Tables

General

1.1 Introduction

With the increase of ship size and speed, shipboard vibration becomes a greater concern in the design and construction of oceangoing vessels. Excessive ship vibration is to be avoided for passenger comfort and crew habitability. In addition to the undesirable effects on humans, excessive ship vibration may result in the fatigue failure of local structural members or malfunction of machinery and equipment. The guidance provided in this book is designed to provide users, specifically shipyards, naval architects, and ship owners, with practical guidance on the concept design to avoid excessive ship vibration at an early design stage. If simple procedures are followed with insight and good judgment in the concept design stage, then the difficult countermeasures and corrections at the subsequent design stages may be avoided in most cases. This guidance also assists with the finite element analysis (FEA) based vibration analysis procedure to predict the vibration response and evaluate the design in detail design stage. The vibration analysis procedure represents the most current analysis practice used by the majority of classification societies. Subsequently, this book also provides high level guidelines on the vibration measurement procedure during the sea trials and the acceptance criteria on vibration limits based on international standards and classification society practice.

1.2 Application and Scope

This guidance is applicable to the vessels of all lengths, including ship-type floating production units (FPU). It is designed to provide preliminary reference on ship vibration excited by the main engine, propeller, or slamming. Throughout this book, the following subject areas are considered:

© The Author(s), under exclusive license to Springer Nature Switzerland AG 2025
F. Karkori, *Ship Vibration 1*, Synthesis Lectures on Ocean Systems Engineering,
https://doi.org/10.1007/978-3-031-75072-4_1

(1) Concept design,
(2) Vibration analysis,
(3) Measurements, and
(4) Acceptance criteria.

The concept design in Chaps. 2, 3 and 4 provides users with immediate, direct, and concise guidance in effectively dealing with ship vibration in the concept design stage. In attempting to provide a sound and concise single point of reference, this book only identifies the most serious problem areas that have caused difficulties to the maritime and shipbuilding industry, and therefore concentrates on those areas. In the concept design, local vibration is not addressed because detail information is not usually available in the early design stage. Instead, the concept design is focused on those areas that have been known to be of critical importance in avoiding harmful ship vibration. The vibration analysis in Chap. 5 provides the FE-based vibration analysis procedure based on first principles direct calculations. The FE-based vibration analysis is recommended to evaluate the design during the detail design stage. If found necessary, the local vibration is to be addressed in the detail vibration analysis. The analysis procedure provides guidelines on FE modelling, engine and propeller excitation, and free and forced vibration analysis. The procedure for calculating slamming excitation can be found in Chap. 10 through Chap. 18. Site-specific wave scatter diagrams should be used for slamming calculations.

For the assessment of ship vibration performance, the actual vibration levels at the most critical locations are to be measured and evaluated during the sea trials. Chapter 6 provides guidelines on the vibration measurement procedure on the instrumentation, measurement conditions and locations, data processing and reporting. Chapter 7 provides acceptance criteria on the vibration limits for human comfort and habitability, local structures and machinery based on international standards and classification society practice. The shaft alignment and torsional vibration are not directly addressed in this book. The overall procedure for ship vibration assessment as recommended herein is shown at Fig. 1.1.

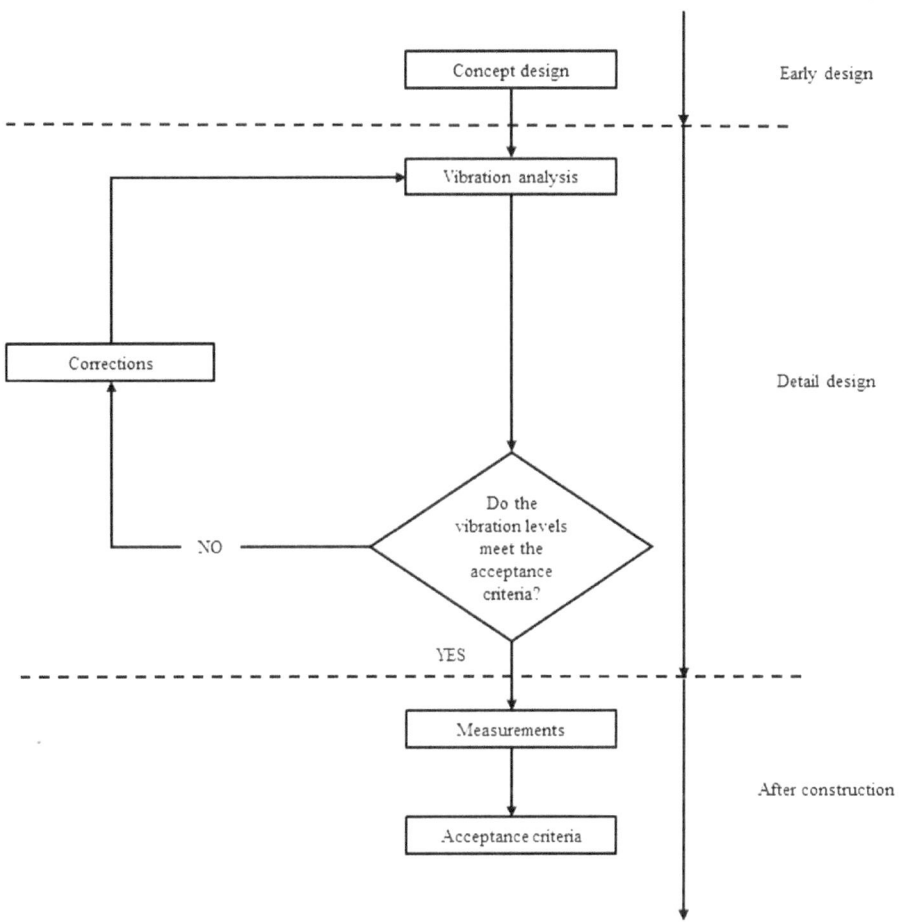

Fig. 1.1 Overall procedure for ship vibration assessment

2.1   Introduction

Concept design is where the vibration avoidance process must begin. It is clear that if the vibration problems, repeatedly identified by experience as the most important, are addressed at the earliest design stage, ultimately serious problems, involving great cost in correction efforts, may be avoided. The focus is on planning for vibration early at the Concept Design stage, where there has been no development of details. If as much as possible can be done in concept design with the simple tools and rules of thumb available at that level, it will help to avoid major vibration problems. The major potential problems may often be present in the crude concept design definition. Just identifying and addressing those potential problems in terms of the minimal technology available at the concept design stage is considered very important to the success of ship design. In this chapter through Chap. 4 provides guidelines on the concept design. Some of these guidelines are presented in the publication *Principles of Naval Architecture*, Chap. 7 (SNAME, 1988).

2.2   Design Considerations

The four elements of importance in ship vibration are:

- Excitation,
- Stiffness,
- Frequency ratio, and
- Damping.

It should be noted that any of the following contribute to vibration reduction:

F. Karkori, *Ship Vibration 1*, Synthesis Lectures on Ocean Systems Engineering,
https://doi.org/10.1007/978-3-031-75072-4_2

(1) *Reduce exciting force amplitude, F.* In propeller-induced ship vibration, the excitation may be reduced by changing the propeller unsteady hydrodynamics. This may involve lines or clearance changes to reduce the non-uniformity of the wake inflow or may involve geometric changes to the propeller itself. Specifics in this regard are addressed in Chap. 3.

(2) *Increase stiffness, K.* Stiffness is defined as spring force per unit deflection. In general, stiffness is to be increased rather than decreased when variations in natural frequency are to be accomplished by variations in stiffness. It is not a recommended practice to reduce system stiffness in attempts to reduce vibration.

(3) *Avoid values of frequency ratio near unity;* $\frac{\omega}{\omega_n} = 1$ is the resonant condition. At resonance, the excitation is opposed only by damping. Note that $\frac{\omega}{\omega_n}$ can be varied by varying either excitation frequency ω or natural frequency ω_n. The spectrum of ω can be changed by changing the RPM of a relevant rotating machinery source, or, in the case of propeller-induced vibration, by changing the propeller RPM or its number of blades. ω_n is changed by changes in system mass and/or stiffness; increasing stiffness is the usual and preferred approach. Specific measures for resonance avoidance in ships are addressed in Chap. 4.

(4) *Increase damping, ζ.* Damping of structural systems in general, and of ships in particular, is small; $\zeta < < 1$. Therefore, except very near resonance, the vibratory amplitude is approximately damping independent. Furthermore, damping is difficult to increase significantly in systems such as ships; ζ is, in general, the least effective of the four parameters available to the designer for implementing changes in ship vibratory characteristics.

2.3   Concept Design Approach

Four elements were identified in the preceding as being influential in determining ship vibratory response, and their relationship to vibration reduction was addressed. While quantification of all four elements is required in calculating the vibration response level, acceptable results may consistently be achieved with reasonable effort by focusing attention in concept design on two of the four elements. The two of the four elements of importance are excitation and frequency ratio. The achievement in design of two objectives with regard to these elements has resulted in many successful ships:

- Minimise dominant vibratory excitations, within the normal constraints imposed by other design variables, and
- Avoid resonances involving active participation of major subsystems in frequency ranges where the dominant excitations are strongest.

Unlike vibration response, the excitation and frequency ratio elements involved in these objectives may be predictable with useful reliability for making design judgments. For the most troublesome case of propeller excitation, detailed hydrodynamic calculation procedures in conjunction with model testing have been established, at least to the level of reliable relative predictions in the post-concept, preliminary design stage. Natural frequencies involving the ship hull and its major subsystems are usually predictable using judicious modelling and modern numerical structural analysis methods. Accuracy levels achievable in predictions of propeller and engine excitation and of sub-system natural frequencies are generally considered to be reliable to consistently achieve the two objectives cited above. The detailed calculations and experiments required to avoid resonance and to minimise excitation are usually performed by engineering analysis groups or model basins and are usually not the immediate responsibility of the ship designer. If addressed early enough while basic changes in a design can still be made, excessive vibration may be avoided effectively. The main functions of the ship designer in this regard are:

- To establish a concept design to serve as the subject of the detailed investigations, and
- To judge whether detailed analysis is warranted as a further step.

The quality of the concept design will be reflected in the number of detailed iterations, if any, that are required for achieving an acceptable final design. For the purpose of establishing high-quality concept designs, which may require time consuming calculations and model testing, the designer needs both guidance as to the areas of the design likely to be in most need of attention, and some simple methodology for identifying the critical aspects. Experience has shown that attention to vibration in concept design of large ships can usually be paid to the following items:

(1) Hull girder vertical vibration excited by the main engine,
(2) Main machinery/shafting system longitudinal vibration excited by the propeller, and
(3) Superstructure fore-and-aft vibration excited by hull girder vertical vibration and/or main propulsion machinery/shafting system longitudinal vibration.

A myriad of local vibrations, such as hand-rails, antennas, plating panels, etc., may be encountered on new vessel trials in addition to these three. But local problems usually involve local structural resonances and often considered as minor problems, as the correction approach by local stiffening may be easily achievable. Chapters 3 and 4 provide general guidance on the methodology of established effectiveness in dealing with the three critical items cited above in the concept design stage. The concept design checklist is summarised in Chap. 9, Sect. 2.2. Below are the important areas to be considered during concept design:

(1) Main engine excitation (refer to Chap. 3, Sect. 2.3),
(2) Stern lines and propeller clearance (refer to Chap. 3, Sect. 5),
(3) Alternating thrust and cavitation (refer to Chap. 3, Sect. 7),
(4) Hull girder vertical vibration (refer to Chap. 4, Sect. 2.3),
(5) Machinery/shafting longitudinal vibration (refer to Chap. 4, Sect. 5), and
(6) Super structure fore-and-aft vibration (refer to Chap. 4, Sect. 7).

3

3.1 Introduction

It is appropriate that the principal vibration exciting sources be addressed first, since with high excitation levels excessive vibration can occur almost independently of system structural characteristics. In general, the major sources are the low-speed diesel main engine and the propeller. Gas turbines are generally considered to give less excitation than diesel engines. Thus, in this chapter, attention is paid to the excitation of the low-speed main diesel engine.

3.2 Low-Speed Main Diesel Engine

Significant progress has been made in recent years by engine manufacturers in reducing vibratory excitation, largely by moment compensators installed with the engine. Steps to be taken by the engine manufacturer are to be addressed in main engine specifications for a new vessel. In this regard, it is important for the shipyard or owner to be technically knowledgeable on this issue. Diesel engine vibratory excitation can be considered as composed of three periodic force components and three periodic moment components acting on the engine foundation. Actually, the periodic force component along the axis of the engine is inherently zero, and some other components usually balance to zero depending on particular engine characteristics. Two distinctly different types of forces can be associated with the internal combustion reciprocating engine. These are: (a) gas pressure forces due to the combustion processes (guide force couples) and (b) inertia forces produced by the accelerations of the reciprocating and rotating engine parts (external forces). Figure 3.1 shows the typical external force and moments acting on a diesel engine.

© The Author(s), under exclusive license to Springer Nature Switzerland AG 2025
F. Karkori, *Ship Vibration 1*, Synthesis Lectures on Ocean Systems Engineering,
https://doi.org/10.1007/978-3-031-75072-4_3

Fig. 3.1 External forces and
moments

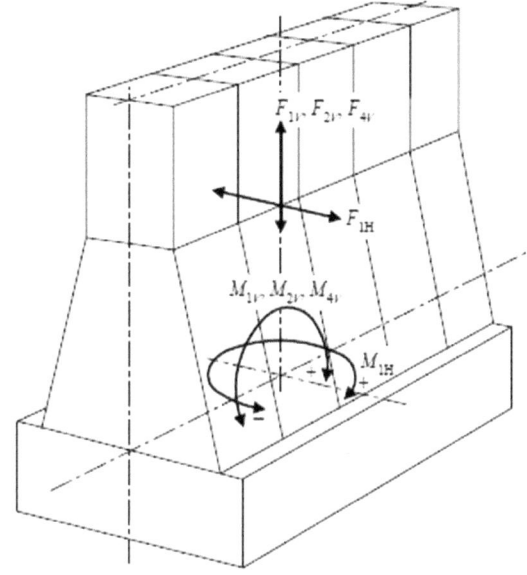

Guide force couples acting on the crosshead result from transverse reaction forces depending on the number of cylinders and firing order. The guide force couples cause rocking (H-couples) and twisting (X-couples) of the engine, as shown in Fig. 3.2. The engine lateral vibration due to the guide force couples may cause resonance with the engine foundation structure. A possible solution at concept design stage may be a consideration of lateral stays (top bracings), connecting the engine's top structure to the ship hull.

The vertical force and moment, which are of primary concern with regard to hull vibratory excitation, and the transverse force and moment as well, are due to unbalanced inertial effects. For engines of more than two cylinders, which is the case of interest with ships, the vertical and transverse inertia force components generally balance to zero at the engine foundation. This leaves the vertical and transverse moments about which to be concerned. The values of the moment amplitudes are usually tabulated in the manufacturer's specification for a particular engine. The majority of low-speed marine diesels currently in service have six cylinders or more. Therefore, the second order vertical moment M_{2v} is generally considered to contribute the most to the hull vibration. However, depending on the specific number of cylinders, the first order or higher order moment can be as large as the second order moment. In that case, further consideration is to be given to the first or higher order moment.

A hull girder mode up to the third or fourth can have a natural frequency as high as the twice-per revolution excitation of the second order vertical moment. Hull girder modes higher than the first three or four have diminishing excitability and may be of less concern. The following steps are therefore recommended in concept design:

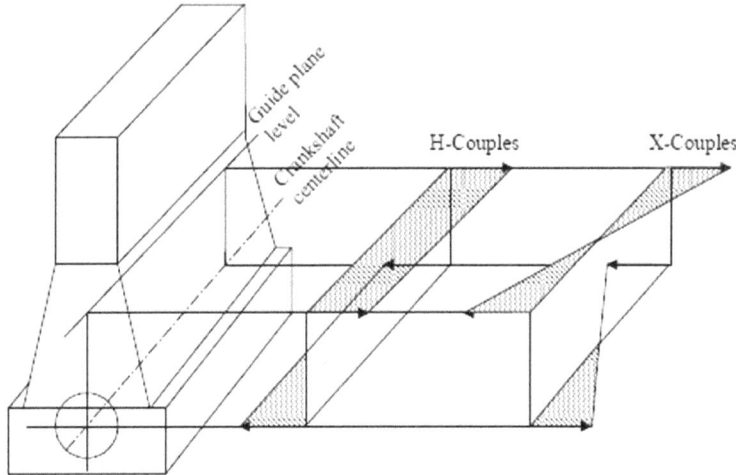

Fig. 3.2 Guide force couples

(1) The second order vertical moment M_{2v} is the diesel engine excitation of most concern. The potential danger is in resonating one of the lower hull girder vertical modes with a large second order vertical moment. The value of M_{2v} is to be requested from the potential main engine manufacturer, as early as possible.
(2) Power related unbalance (PRU) values may be used to determine the acceptable level of M_{2v}

$$PRU = \frac{M_{2v}[N - m]}{Engine\ Power[kW]}$$

Further attention is recommended in cases where PRU exceeds 220 N-m/kW. The action recommended at the initial engine selection stage may be either change of engine selection or installation of moment compensators supplied by the engine manufacturer. Otherwise, the vertical hull girder response is to be checked by calculation within an acceptable level without installation of compensators.
(3) The engine lateral vibration due to X-type and H-type moments may produce excessive local vibration in the engine room bottom structure depending on the engine frame stiffness and engine mounting. The installation of lateral stays on the engine room structure is to be addressed at the early design stage.

3.3   Hull Wake

Hull wake is one of the most critical aspects in avoidance of unacceptable ship vibration. Propeller-induced vibration problems in general start with unfavourable hull lines
in the stern aperture region, as manifest in the non-uniform wake in which the propeller
must operate. Unfortunately, propeller excitation is far more difficult to quantify than
the excitation from internal machinery sources. This is because of the complexity of
the unsteady hydrodynamics of the propeller operating in the non-uniform hull wake. In
fact, the nonuniform hull wake is the most complicated part; it is unfortunate that it is
also the most important part. Propeller-induced vibration would not be a consideration in
ship design if the propeller disk inflow were circumferentially uniform. Any treatment of
propeller excitation must begin with a consideration of the hull wake.

For engineering simplification, the basic concepts allow for the circumferential non-
uniformity of hull wakes, but assume, for steady operation, that wake is time invariant
in a ship-fixed coordinate system. Nominal wake data from model scale measurements in
towing tanks are presented either as contour plots or as curves of velocity versus angular
position at different radii in the propeller disc. Figure 3.3 shows the axial and tangential
velocity components for a typical conventional stern merchant ship.

The position angle, θ, on Fig. 3.3 is taken as positive counterclockwise, looking forward, and x is positive aft. The axial wake velocity v_X and tangential wake velocity v_T
are dimensionless on ship forward speed, U. Note that the axial velocity is symmetric in
θ about top-dead-centre (even function) and the tangential velocity is asymmetric (odd
function). This is a characteristic of single screw ships due to the transverse symmetry
of the hull relative to the propeller disk; such symmetry in the wake does not, of course,
exist with twin-screw ships.

The wake illustrated above represents one of the two characteristically different types
of ship wakes. The flow character of the conventional skeg-stern is basically waterline
flow; the streamlines are more or less horizontal along the skeg and into the propeller
disk. The flow components along the steep buttock lines forward of the propeller disk
are small. The dominant axial velocity field of the resultant wake has a substantial defect
running vertically through the disk along its vertical centreline, at all radii.

This defect is the shadow of the skeg immediately forward. The tangential flow in the
propeller disk, being the combination of the component of the upward flow toward the free
surface and any disk inclination relative to the baseline, is much smaller. The idealisation
of this wake, for conceptual purpose, is the two-dimensional flow behind a deep vertical
strut placed ahead of the propeller. In this idealisation, the axial velocity distribution
is invariant vertically, and any tangential velocity distribution (due to disk inclination)
is asymmetric about the vertical disk axis. This basic characteristic is exhibited in the
Fig. 3.3.

A characteristically different wake flow is associated with the strut or barge-type stern,
the upper of Fig. 3.4, which has a broad counter above the propeller disk and minimal

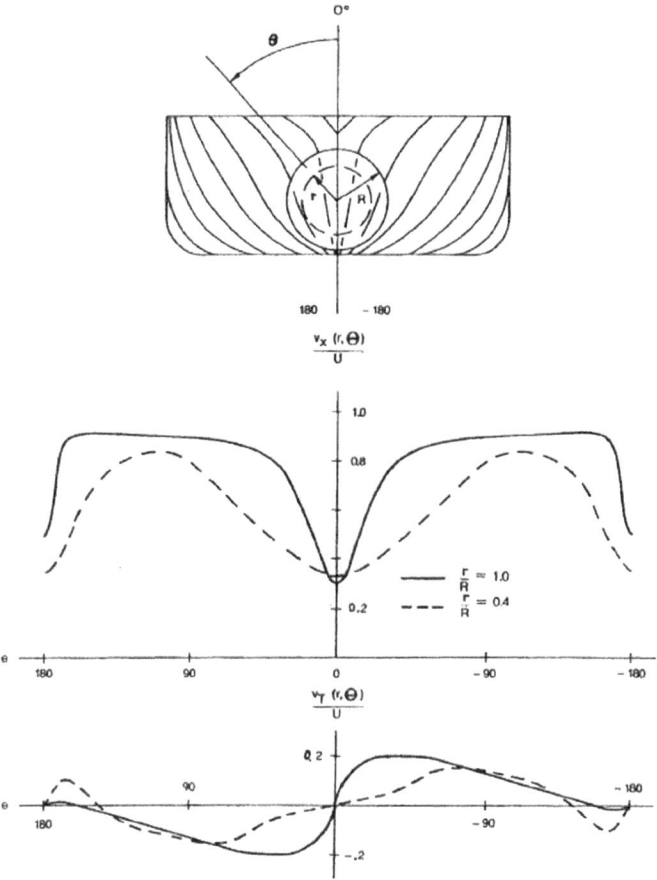

Fig. 3.3 Nominal wake distribution for a typical merchant ship (DTMB Model 4370, $C_B = 0.6$)

irregularity immediately forward. The engine is further forward with this stern-type to accommodate finer stern lines needed to minimise wave resistance in high-speed ships, although an open strut stern would be beneficial for vibration minimisation at any speed (provided the buttocks lines are not too steep). The flow character over this type of stern is basically along the buttock lines, versus the waterlines. Some wake nonuniformity may be produced by appendages forward, such as struts and bearings or by shaft inclination, but the main wake defect, depending on the relative disk position, will be that of the counter boundary layer overhead. In this case, assuming minimal shaft inclination, substantial axial wake again exists, but only in the top of the disk.

Generally, only the blade tips penetrate the overhead boundary layer, and the axial wake defect occurs only at the extreme radii near top-dead-centre, rather than at all radii along the vertical centreline, as in the case of the conventional single screw stern. Just

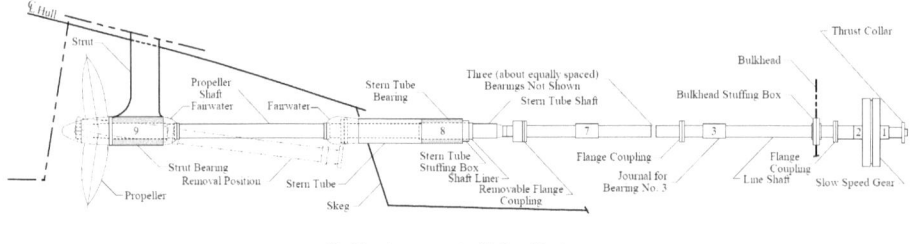

Shafting Arrangement with Strut Bearing

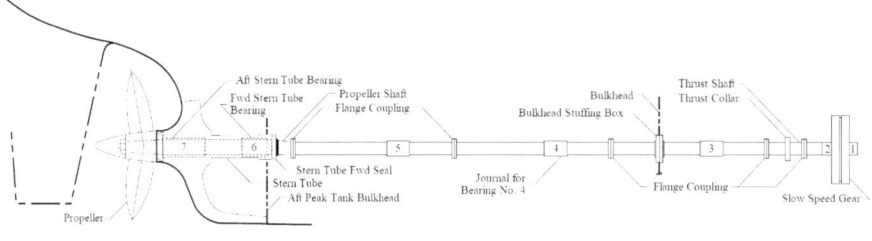

Shafting Arrangement without Strut Bearing

Fig. 3.4 Alternative shafting arrangements: open strut stern (upper); conventional skeg stern (lower)

as in the case of the conventional stern, the tangential disk velocity with the strut stern will generally be small; the vertically upward velocity ratio through the propeller disk will have average values on the order of the tangent of the sum of the buttock and shaft inclination angles. The idealisation in the case of the barge stern, as a sequel to the vertical strut idealisation of the wake of the conventional stern, is a horizontal flat plate above the propeller. Here, the degree of axial wake non-uniformity depends on the overlap between the propeller disk and the plate boundary layer. The tangential (and radial) wake components are due entirely to the shaft relative inclination angle in this idealisation, as the flat plate boundary layer produces only an axial defect.

3.3.1   Hull-Propeller Clearance

The distinction between the two basically different wake types is useful in understanding the importance of clearances between the propeller blades and local hull surfaces. First of all, it is helpful to consider the hull surface excitation by the propeller as composed of two effects:

(1) *Wake effect.* The effect of changing the wake inflow to the propeller according to a specified propeller relocation, but with the propeller actually fixed in position relative to the hull, and

(2) *Diffraction effect.* The effect of changing the propeller location relative to the hull, but with the wake inflow to the propeller held fixed.

It is a common misconception that the cruciality of propeller-hull clearances has to do primarily with the diffraction effect. To the contrary, analysis shows that for wake inflow held invariant, propeller-induced excitation level is relatively insensitive to near-field variations in propeller location. It is the high sensitivity of propeller blade pressure and cavitation inception to the variations in wake non-uniformity accompanying clearance changes that dictates the need for clearance minima. In general, the wake gradients become more extreme as propeller-hull clearances are decreased.

The conventional and strut-stern wake types are to be considered in light of the above fact. It is probable that too much emphasis is often placed on aperture clearances in conventional stern single-screw ships. From the point of view of vertical clearance, there is no significant boundary layer on the usually narrow counter above the propeller with this stern type. Furthermore, from the point of view of the vertical strut idealisation of the skeg, the axial velocity distribution would be invariant with vertical disk position. The critical item with vertical tip clearance in the conventional stern case seems to be the waterline slope in the upper skeg region. Blunt waterline endings can result in local separation and substantially more severe wake gradients in the upper disk than suggested by the simple strut idealisation; a "blunt strut" idealisation then would be more appropriate. Fore-and-aft clearances in the conventional stern case are generally less critical than the vertical clearances. Wakes attenuate very slowly with distance downstream. While increasing the fore-and-aft clearances between the blade tips and the skeg edge forward certainly acts to reduce the wake severity, the reduction will be marginally detectable within the usual limits of such clearance variation. An exception would exist in the case of separation in the upper disk due to local waterline bluntness. The closure region of the separation cavity in that case exhibits large gradients in axial velocity.

On the other hand, for the broad and flat countered strut stern vessel, the vertical tip clearance is a much more critical consideration. A relatively uniform wake will result if the propeller disk does not overlap the overhead boundary layer and the shaft inclination is moderate. This is the condition, in general, achieved on naval combatant vessels; the usual practice in naval design is a minimum vertical tip clearance of one quarter propeller diameter. Vibration problems are almost unheard of on naval combatants.

Some wake non-uniformity on strut stern ships results from shaft struts and from the relatively high shaft inclinations often required to maintain the 25 percent overhead tip clearances. With proper alignment to the flow, shaft struts produce highly localised irregularities in the wake which are generally not effective in the production of vibratory excitation. The main effect of shaft inclination is a relative up-flow through the propeller

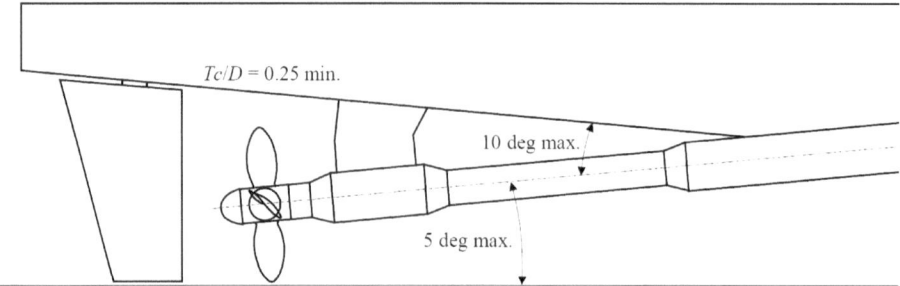

Fig. 3.5 Open strut stern arrangement

disk, although separated flow from forward off of a highly inclined propeller shaft can produce a serious axial defect. The cavitation that can result at the 3 and 9 o'clock blade positions has shown to be of concern with regard to noise and minor blade erosion, but the hull vibratory excitation produced has not been found to be of much significance for strut stern ships.

The minimum vertical tip clearance of 25 percent of a propeller diameter is more or less accepted as the standard in commercial practice as well as naval. Consistent with the preceding analysis, the following lists the recommended configuration for stern arrangements in the order of preference for avoidance of excessive propeller induced vibration:

(1) Single/twin screw strut stern
   • The minimum vertical tip clearance is not to be less than 25 s of the propeller diameter,
   • The shaft inclination angle relative to the baseline is not to be more than 5 degrees,
   • The shaft inclination angle relative to the buttocks angle of the counter is not to be more than 10 degrees. Refer to Fig. 3.5.
(2) Conventional stern with skeg and bossing
   • The waterline angle from the vertical centre plane at the entrance to the aperture just forward of the top of the propeller disk is not to exceed 35 degrees. Refer to Fig. 3.6.
   • With regard to the propeller tip clearance, the conventional skeg-stern ships are less critical than the strut-stern ships, as discussed earlier. In commercial practice, a minimum vertical tip clearance on order of 25% of propeller diameter and forward clearance of 40% of propeller diameter are often employed as a usual practice.
(3) In the event that these limits cannot be achieved in concept design, it is recommended that the decision be made to proceed on to model testing and/or direct calculation for confirming or establishing stern lines.

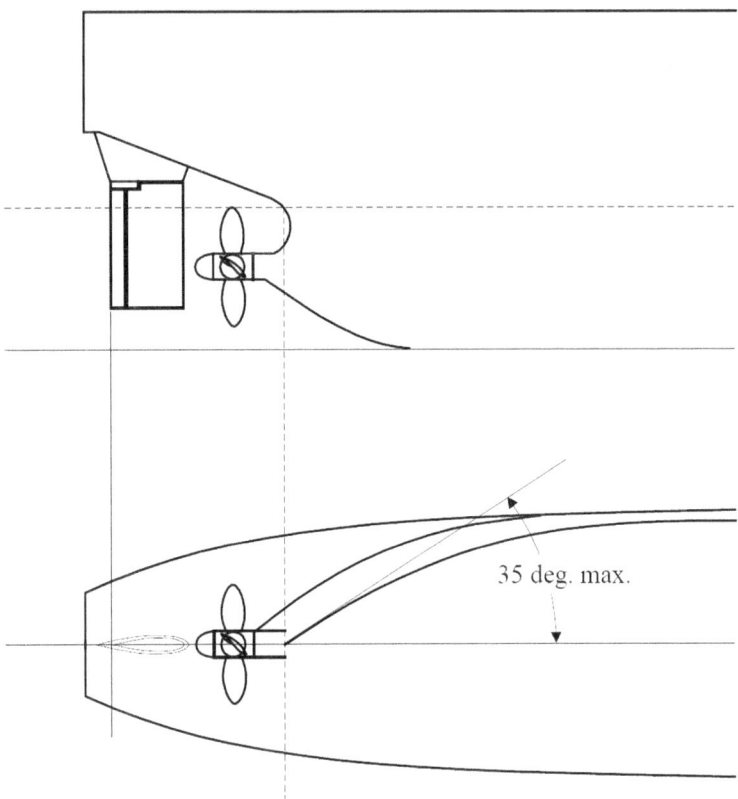

Fig. 3.6 Conventional skeg-stern arrangement

3.4   Propeller

With an unfavourable wake, propeller compromises are then usually required to achieve compensation. Two types of excitations are of primary concern in conjunction with the three main items identified as critical in Chap. 2, Sect. 5.

(1) Alternating thrust exciting longitudinal vibration of the shafting and machinery, and
(2) Vertical pressure forces on the stern counter exciting hull and superstructure vibration.

3.4.1   Alternating Thrust

Alternating thrust, the excitation for longitudinal vibration of the shafting/main machinery system, occurs at blade rate frequency (Propeller RPM × Blade number N) and its multiples. The fundamental is usually much larger than any of its harmonics, however. Alternating thrust is produced by the blade number circumferential harmonic of the hull wake. This suggests that the higher the blade number the better, since the wake harmonic series does converge. However, around the typical blade number of 4, 5, and 6, the wake harmonic series convergence is not well organised, and is not a primary consideration. In fact, with the wake of a conventional single-screw stern, lower alternating thrust favours an odd-bladed propeller. This is because of, for an even bladed propeller, the line-up of opposite blades with the characteristic wake spike along the vertical centre plane above and below the propeller axis. This wake characteristic was discussed in Sect. 3.3, on hull wake. With strut-stern ships there is typically little blade number bias on the basis of wake.

Misconceptions exist about the effectiveness of propeller blade "skew" in reducing vibratory excitation. Skew is the tangential "wrapping" of the blades with radius. Positive skew is in the angular direction opposite to the direction of rotation. For the case of the conventional single screw ship wake, it has been discussed that the shadow of the vessel skeg produces a heavy axial wake defect concentrated along the disk vertical centreline. The blades of conventional propellers ray-out from the hub (i.e., the blade mid chord lines are more or less straight rays emanating from the hub centreline). Such "unskewed" blades abruptly encounter the axial velocity defect of the conventional stern wake at the top and bottom-dead-centre blade positions. The radially in-phase character of the abrupt encounter results in high net blade loads and radiated pressure.

A more gradual progression of the blades through the vertical wake defect is accomplished by curving the blades. Different radii enter and leave the wake spike at different times; cancellation results in the radial integrations to blade loads and radiated pressure, with the result of potentially significantly reduced vibratory excitation.

Percentage skew is the blade tip skew angle, relative to the blade ray through the mid-chord of the propeller hub section, divided by the blade spacing angle (e.g., for 5 blades the blade spacing angle is 72 degrees; therefore, a tip skew angle of 72 degrees would be 100% skew). Refer to Fig. 3.7, where ray 'A' is passing through the tip of blade at mid-chord line and ray 'B' is tangent to the mid-chord line on the projected blade outline.

Highly skewed propellers may be designed to have efficiency equivalent to propellers with minimal skew, but there have been problems with blade strength and flexibility. The modern trend has been a movement back toward more moderate positive tip skew, but with negative skew in the blade root regions. Here the percent skew may be defined in terms the total angle enclosed by the angular maxima of the blade leading and trailing edges. This provides blades with more balanced skew distributions, and highly *swept* leading edges. The radial load cancellations are achieved as effectively, and blade strength is less problematic. The balanced skew is beneficial to limiting blade spindle torque with controllable pitch propellers and is used widely with naval combatant vessels.

Skew will inherently work less effectively with strut-stern wakes since the axial velocity defect tends to be concentrated more in the outer extreme radii. The more radially

Fig. 3.7 Maximum skew angle

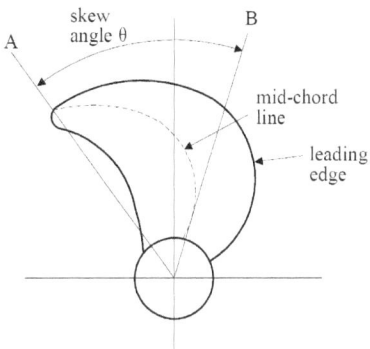

uniform distributions of the conventional stern case are not available with which to achieve as high a degree of *dephasing* and radial cancellation. Of course, the strut stern vessel is in less need of propeller design extremes, as vibration problems are already largely eliminated by the stern form selection, provided proper clearances are incorporated. Care must be taken in incorporating skew, particularly in replacement propellers, that adequate clearances between the blade tips and the rudder be maintained. As the blades are skewed in the pitch helix the tips move downstream, closing-up blade tip to rudder clearances. The consequences can be increased hull vibratory forces transmitted through the rudder, as well as rudder erosion caused by the collapsing sheet cavitation shed downstream off the blade tips as they sweep through the top of the propeller disk. The recourse is to incorporate warp into the blades along with the skew. Warping is a forward raking of the skewed blades back to (and sometimes beyond) the propeller disk. It is equivalent to skewing the blades in the plane of the disk rather than in the pitch helix.

It is noted that skew (and/or sweep) has a beneficial effect in reducing the effects of vibration-producing unsteady sheet cavitation, even when such cavitation may be concentrated in the blade tips. The blade curvature is thought to result in a radially outward flow component in the vicinity of the blade tips which tends to sweep the cavity sheets into the tip vortex, where the critical collapse phase occurs more gradually. Unsteady sheet cavitation is found to have minimal effect on alternating thrust but can dramatically amplify the hull pressure force components. This is addressed specifically in this chapter, Sect. 4.2. If the guidance offered on stern lines with regard to hull wake are adopted, then moderate and acceptable alternating thrust will result with propellers properly sized for propulsion considerations. This is without the need, particularly at the concept design stage, for detailed analysis to quantify alternating thrust level further along. Such quantification for detailed analysis is considered in Chap. 5.

Some ranges of the blade-rate alternating thrust amplitude are given below to serve as
a reference for engineering decision making at the concept design level. The values are
given as percentages of steady ahead thrust, and *assume that due care has been taken with
stern lines and that the propeller is properly sized for proper propulsive performance:*

	Alt thrust/steady thrust
Conventional stern, moderately skewed/swept	0.02–0.06
Strut stern	0.005–0.03

More details on propeller bearing forces and moments including alternating thrust for
20 real ship cases are given in Chap. 5. There is an outstanding issue here regarding
proper sizing of the propeller, which is needed in the trade-offs at the concept design
level. The old Burrill (1943) cavitation criterion is still useful in this regard. As stated,
alternating thrust is not very sensitive to blade cavitation, but limiting the steady cavitation
to accepted levels assures that the propeller is not overloaded from propulsive considerations, which is recommended as necessary to achieve the alternating thrust ranges listed
above. The Burrill chart is at Fig. 3.8, which is a plot of a thrust loading coefficient at the
0.7 propeller blade radius, τ_c tc, versus the blade cavitation number, σ, also at the 0.7
radius.

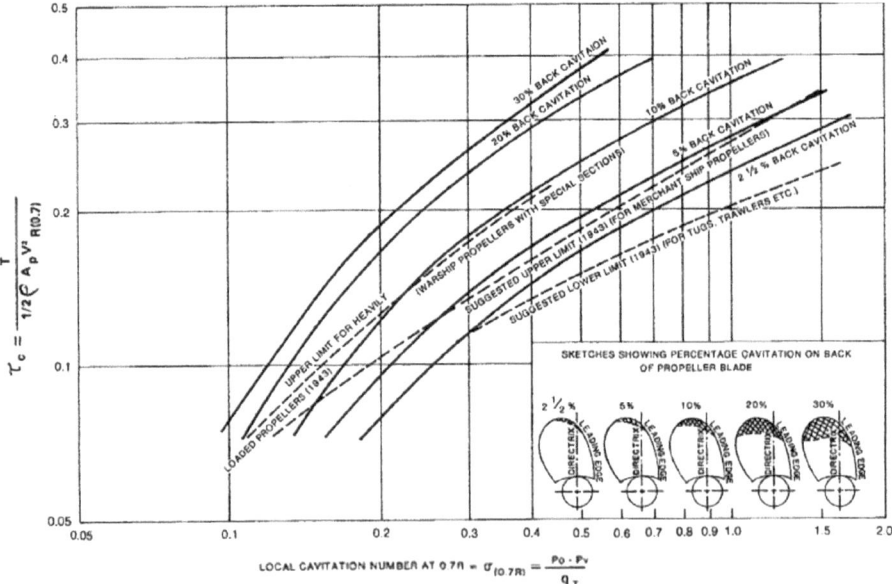

Fig. 3.8 Burrill cavitation inception chart

$$\tau_c = \frac{T}{A_p q_T}$$

and

$$\sigma = \frac{p_a + \rho g h - p_c}{q T}$$

where

T propeller thrust in kN.
A_p propeller projected area in m^2
q_T 12
V_R relative velocity of water at 0.7 radius in m/sec.
ρ_a absolute atmospheric pressure in kPa.
h depth of the propeller axis in m.
p_c cavitation cavity absolute pressure in kPa.
ρ water density in ton/m^3

The relative velocity V_R at 0.7 radius is:

$$V_R(0.7R) = U(1 - w)\sqrt{1 + \left(\frac{0.7\pi}{J}\right)^2}$$

where

U ship speed.
w wake fraction.
J design advance coefficient.

The cavity pressure is often taken to be pure vapor pressure at the ambient water temperature, but for developed cavitation as is the case here, versus inception, pc is higher due to dissolved air in the water; pc around 6–7 kPa is often assumed. The curves of Fig. 3.8 correspond to the degree of steady cavitation development. The 5% back cavitation line was suggested as the limit for merchant ship propellers back in 1943. However, for modern commercial ship propellers, which are universally designed with aerofoil-type blade sections, the 10% back cavitation line is probably more appropriate. But either the 5 or 10% lines, considering the cavitation developments shown, may serve the purpose to avoid overloading the propeller in steady propulsion.

An example of the use of the Burrill chart for this purpose is as follows:

- Take a single screw ship with a delivered power $P_d = 22{,}380$ kW and a max rated speed $U = 22$ knots, corresponding to a thrust, $T = 1{,}400$ kN. The propeller diameter has been set at six metres consistent with tip clearance maxima. The $PRPM = 120$ and the wake fraction is estimated as $w = 0.2$. The pitch ratio at the 0.7 radius, $P/D]_{0.7R} = 0.9$. This gives an advance coefficient, $J = 0.755$ and resulting q_T of 399.3 kPa. For the hub centreline depth $h = 4.5$ m, $p_a = 101.3$ kPa, and $p_c = 6.9$ kPa, $\sigma = 0.35$. If the 5% back cavitation line is selected, $\tau_c \cong 0.16$. With the disk area $A_0 = \frac{\pi D^2}{4}$, the required projected area ratio would be:

$$\frac{A_p}{A_0} = \frac{4T}{\pi \tau_c q_T D^2}$$

- Substituting the values, $A_p/A_0 = 0.775$. It is preferable to work with the developed area ratio, A_d/A_0, where A_d denotes propeller developed area with zero pitch. Taylor's approximate formula gives the projected area in terms of the developed area as:

$$\frac{A_p}{A_d} \cong 1.067 - 0.229\frac{P}{D}]_{0.7R}$$

Then the required developed area ratio is:

$$\frac{A_d}{A_0} = 0.90$$

- The projection here is then, with a good wake, as to be achieved by following the provided recommendations in this book, the subject propeller blades may not be heavily loaded in the mean, and the alternating thrust is to be in approximately direct proportion. As an alternative, if it was decided to use a less conservative approach and accept steady cavitation to the 10% back cavitation at $\sigma = 0.35$, $\tau_c \cong 0.2$. Then the alternative developed area ratio would be:

$$\frac{A_d}{A_0} = 0.72$$

Of course, the more conservative propeller, with the larger blades, will have lower propulsive efficiency. The following is recommended in concept design, in addition to the preceding stern lines recommendations relative to hull wake, for achieving not excessive alternating thrust from the propeller design consideration:

(1) Perform the Burrill cavitation calculation demonstrated above with a blade area ratio as needed to limit the steady cavitation up to 10% back cavitation.

(2) For a conventional stern, a 5-bladed propeller is favoured in reducing the alternating thrust, unless any harmful resonant vibration is anticipated on superstructure, shafting system or local structures.

(3) In general, incorporate blade skew of no more than 50%, or if a blade with a highly swept leading edge is adopted (negative skew in the root), use no more than 25 degrees of tip skew (for either stern type). For a highly skewed propeller design other than the foregoing, refer to the Class Rules for the specific vessel in question.

3.4.2 Hull Pressure Forces

The dominating excitation for ship hull vibration, related to the third of the three critical items identified in the preceding, is propeller cavitation-induced hull surface pressure forces. If intermittent blade cavitation does not occur to a significant degree, then the main excitation of the hull is via the shafting system and the main engine, as already discussed. These sources will cause minor problems relative to, at times, the intense vibration from an intermittently cavitating propeller. It was in the mid-70's that propeller blade unsteady cavitation, triggered by the wake non-uniformity, was found to be the main culprit in most of the ship vibration troubles.

The sheet cavitation expands and collapses on the back of each blade in a repeating fashion, revolution after revolution. The sheet expansion typically commences as the blade enters the region of high wake in the top part of the propeller disk. Collapse occurs on leaving the high-wake region in a violent and unstable fashion, with the final remnants of the sheet typically trailed out behind in the blade tip vortex. The sheet may envelop almost the entire back of the outboard blade sections at its maximum extent. For large ship propellers, sheet average thicknesses are on the order of 10 cm, with maxima on the order of 25 cm occurring near the blade tip just before collapse.

The cavitation, while of catastrophic appearance, is usually not deleterious from the standpoint of ship propulsive performance. The blade continues to lift effectively; the blade suction-side surface pressure is maintained at the cavity pressure where cavitation occurs. The propeller bearing forces (i.e., alternating thrust) may be largely unaffected relative to non-cavitating values for the same reason. The cavitation may or may not be erosive, depending largely on the degree of *cloud cavitation* (a mist of small bubbles) accompanying the sheet dynamics. The devastating appearance of fluctuating sheet cavitation is manifest consistently mainly in the field pressure that it radiates, and the noise and vibration that it thereby produces. The level of hull surface excitation induced by a cavitating propeller can be easily an order of magnitude larger than typical non-cavitating levels. Vertical hull surface forces due to intermittent cavitation typically exceed propeller bearing forces by large amounts.

The cavitation-induced hull surface force is, like the bearing forces, composed of the fundamental blade rate frequency and it harmonics. However, unlike the character of the alternating thrust discussed in the preceding, the harmonics of the cavitating surface force typically converge very slowly. The fundamental, at PRPM times blade number, is usually largest, but twice and three times blade-rate of the same order typically exists, and can resonate structure and generate noise well above the normal excitation frequency range.

Cavitating vibratory force reduction is achievable with propeller design refinements, as sheet cavitation dynamics is sensitive to blade geometric, as well as wake, detail. As previously noted, blade skew can be beneficial, and tip loading can be reduced to reduce cavitation extent and intermittency by local blade tip pitch reduction, as well as changes in blade tip chord distributions. But measures of this type do reduce propeller efficiency and thereby compromise propulsive performance.

The occurrence of propeller blade intermittent cavitation cannot be ignored in attempts to control propeller induced hull vibration. The issue here is again primarily a hull wake issue. With an unfavourable wake, characterised as exhibiting high gradients in the upper disk, or even a separation pocket, the resulting large cavity volume variations can produce hull surface force amplitudes of up to 30–40% of the steady thrust, with significant levels out to several harmonics of blade-rate frequency. But even with a good wake, as described in the preceding, the conventional stern ship will still exhibit a wake defect in the disk over the depth of the skeg, and some intermittent cavitation will be unavoidable. A vertical surface force with a blade-rate amplitude of 15% of steady thrust is probably a conservative reference for a conventional stern ship with a "good wake" as achievable by the stern lines guidance of the preceding section.

With strut stern vessels with ample tip clearances and low relative shaft inclination, intermittent cavitation, or any cavitation at all, might be avoided entirely. This is largely achieved in the case of radiated noise sensitive naval combatant vessels.

Prediction of the cavitating hull forces is much more complicated, with many more factors contributing, than is alternating thrust, which, as stated, is more or less insensitive to cavitation. Cavitating hull force analysis is certainly not an activity for concept design. However, with any conventional stern ship of modern power level, a propeller cavitation/hull pressure assessment program is to be initiated in the preliminary design stage. For the direct calculation of the propeller-induced hull surface force, refer to the vibration analysis procedure described in Chap. 5.

At concept design, which is the scope of interest here, the same analysis as in the preceding section, using the Burrill criteria for 5–10% steady back cavitation, is to be applied to provide reasonable protection against under-sizing the propeller blades from the standpoint of unsteady cavitation. The example of the preceding section on alternating thrust is therefore proposed as applicable here as well (Table 3.1).

Table 3.1 PRU need for
compensator

PRU	Need for compensator
Below 120	Not likely
120–220	Likely
Over 220	Most likely

4.1   Introduction

The three critical items in concept design are identified in Chap. 2. The relevant excitation having been treated in Chap. 3, the three items are addressed in this section, mainly focused on resonance avoidance.

4.2   Hull Girder Vertical Vibration

The vertical beam-like modes of vibration of the hull girders of modern ships may become serious in three respects:

(1) They can be excited to excessive levels by resonances with the dominant low frequency excitations of slow-running diesel main engines,
(2) Vertical vibration of the hull girder in response to propeller excitation is a direct exciter of objectionable fore-and-aft superstructure vibration, and
(3) Vertical vibration of the hull girder excited by slamming can give rise to excessive vibration of various installations on the vessel's deck, especially slender structures such as a flare tower in hydrocarbon production and process systems.

The propeller is generally incapable of exciting the hull girder modes themselves to dangerous levels. This is primarily because the higher hull girder modes whose natural frequencies fall in the range where propeller excitation is significant have low excitability. However, the low-level vertical hull girder vibration that does occur, either directly from the propeller or indirectly via the main shafting thrust bearing, serves as the base excitation for excessive vibration of superstructures and other attached subsystems which

© The Author(s), under exclusive license to Springer Nature Switzerland AG 2025
F. Karkori, *Ship Vibration 1*, Synthesis Lectures on Ocean Systems Engineering,
https://doi.org/10.1007/978-3-031-75072-4_4

are in resonance with the propeller exciting frequencies. This will be addressed in a later section. The natural frequencies corresponding to the two-noded vertical bending modes of conventional ship hulls can be estimated using Kumai's formula:

$$N_{2v} = \varphi \sqrt{\frac{I_v}{\Delta_i L^3}} cpm$$

where

φ 3.584 × 106- for ship-type floating production units
        3.07 × 106—for other vessel types
I_v moment of inertia, in m4
Δ_i virtual displacement, including added mass of water, in tons
     $\left(1.2 + \frac{1}{3} \cdot \frac{B}{T_m}\right)\Delta$
Δ ship displacement, in tons
L length between perpendiculars, in m
B breadth amidships, in m
T_m mean draft, in m

Table 4.1 gives an indication of the accuracy that can be expected from Kumai's formula. The table compares the prediction of the 2-noded vertical hull bending natural frequency by Kumai's formula with the predictions of detailed finite element calculations performed on different ships. The deviation of the natural frequency calculated using the Kumai formula for ship-type floating production units with respect to collected measured or calculated natural frequencies ranges between −8.6 and 14.7%.

The 2-noded hull vertical bending natural frequencies actually lie well below the dangerous exciting frequencies of either typical diesel main engines or propellers, and are therefore of little consequence in these considerations. As will be demonstrated further

Table 4.1 Comparison of 2-node vertical vibration natural frequencies

Ship No.	Type	Size (dwt)	Kumai (Hz)	FE method (Hz)	Dev. (%)
1	Reefer	15,000	1.54	1.51	+2
2	Ro-Ro	49,000	1.49	1.60	−7
3	Ro-Ro	42,000	1.04	0.94	+10
4	Chemical	33,000	1.00	0.93	+8
5	Bulk carrier	73,000	0.63	0.64	−2
6	Container carrier	120,900	0.41	0.49	−17.0
7	Large container carrier	200,000	0.38	0.45	−15
8	VLCC	363,000	0.40	0.46	−12.8

Table 4.2 Flexible base correction factors

Type	f_e/f_∞
A, C	0.625
B	0.602
D	0.751

on, it is hull girder modes with typically a minimum of 4 or 5 nodes that can be excited excessively by the diesel main engine. In the case of the propeller, the hull girder vertical bending modes that fall near full-power propeller blade-rate excitation are typically more than 7-noded. Full-power blade rate excitation of large ships is usually in the range of 8 to 12 Hz. As indicated in Table 4.2, the 2-noded vertical hull bending mode, on the order of 1 to 2 Hz, is well below the blade-rate excitation frequency level during normal operation.

It is observed that hull girder natural frequencies increase more or less linearly with node number from the 2-noded value for the first few modes. The data shown on Fig. 4.1, from Johannessen and Skaar (1980), provide estimates of the natural frequencies of the first four vertical bending modes of general cargo ships and of the first five vertical bending modes of tankers. Note the good agreement for 2-noded cases. Also note that the 6 Hz maximums represented by Fig. 4.1 still lie well below typical full power propeller excitation frequencies, and the accuracy of the data fits indicated on the figure is deteriorating rapidly as modal order increases. The primary reason for the increasing data scatter with node number is the increasing influence of local effects (i.e., approaching resonances of deckhouses, decks, etc.) on the basic beam modes still identifiable.

The Kumai's formula, in conjunction with Fig. 4.1, is, however, useful in preliminary steps to avoid resonances with a main diesel engine. The following formula, from Johannessen and Skaar (1980), representing the Fig. 4.1 data, expresses the first few vertical bending natural frequencies in terms of the 2-noded value:

$$N_{nv} \approx N_{2v}(n-1)^\alpha$$

where

$\alpha =$ 0.845   General Cargo Ships

        1.0   Bulk Carriers

        1.02   Tankers

Here N_{2v} is the 2-noded vertical natural frequency and n is the number of nodes; n is not to exceed five or six in order to remain within the range of reasonable validity. Note the approximate proportionality of N_{nv} to node number; this is also evident in Fig. 4.1. The use of this data is demonstrated by example:

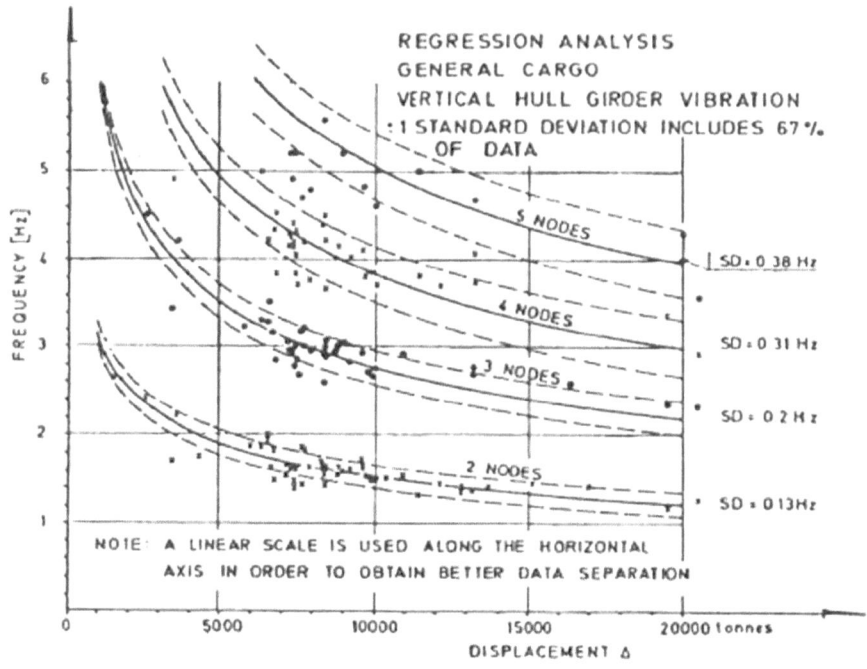

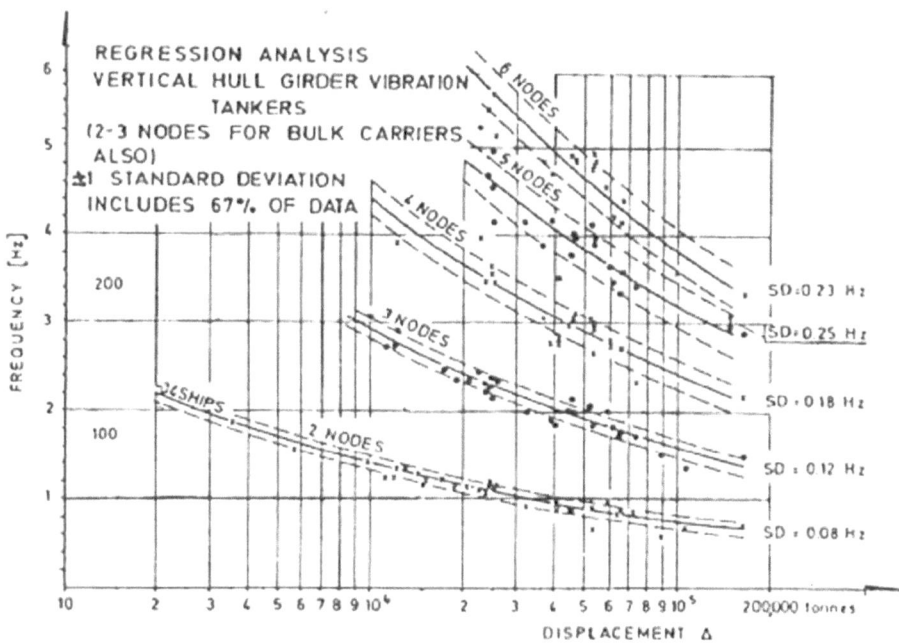

Fig. 4.1   Natural frequencies of vertical hull vibration

- Assuming the "reefer," ship number 1 of Table 4.1, for example, $N_{2v} = 92.4$ cpm, and $\alpha = 0.845$. The first four vertical hull-girder modes of this vessel would then be predicted to have corresponding natural frequencies of:

n	N (cpm)
2	92.4
3	166
4	234
5	298

With the main engine RPM of 122 cpm, for example, the frequency of the second order engine excitation is $2 \times \text{RPM} = 244$ cpm. It is relatively close to the predicted natural frequency of the 4-noded third hull girder mode. While the natural frequency estimate of 234 cpm is indeed rough, it provides at least useful guideline to dictate further analysis to refine the hull girder natural frequency estimates in this particular example.

In the case of projected high excitability in resonant vibration with the diesel engine moments, which does develop in the course of design on occasion, the excitation moment components can usually be reduced effectively by the incorporation of compensators or electric balancers. These devices consist of rotating counterweights usually geared directly to the engine, or electrically powered and installed at aft end of the ship. They are rotated at the proper rate and with the proper phase to produce cancellation with the undesirable first or second order engine generated moment.

An alternative that has seen some popularity is the installation of main diesel engines on resilient mounts (Schlottmann et al. 1999). Isolating the main engine on resilient mounts can be a good approach to minimising hull vibration and structure-borne noise. The following steps are therefore recommended in concept design:

(1) Assuming no moment compensation on the engine, compensators geared for $2 \times$ Engine RPM, is to be installed on both ends of the crank if:
   (a) The full power second-order vertical moment amplitude exceeds the PRU (Power Related Unbalance) value of 220 N-m/kW, as discussed in Chap. 3, Sect. 4.3.
   (b) Twice RPM of the engine at full power is within 20% of any of the vertical hull modes through at least 5-noded as predicted by the procedure of the immediately preceding example.
(2) Moment compensation is effective but should both (a) and (b) occur together, a more precise analysis of the hull girder natural frequencies is recommended in setting struc- ture in preliminary design to more accurately assess the proximity of a resonance with the engine within a 20% band around twice full power engine RPM.

4.3   Main Machinery/Shafting System Longitudinal Vibration Excited by the Propeller

Shafting/machinery longitudinal vibration can also be serious in several respects:

(1) In a resonant or near resonant condition with the system mass and stiffness, thrust reversals at the main thrust bearing can result which, over a relatively short time period, are capable of destroying the thrust bearing,
(2) Engine room vibration, including vibration of the engine itself, can be excessive with regard to foundation and inner-bottom structural distress, and
(3) The amplified thrust transmitted through the main thrust bearing and its moment arm relative the hull girder neutral axis can produce vertical response of the hull girder which excites resonant vibration of hull-mounted substructures.

Interest in longitudinal machinery vibration has a long history, starting seriously with the steam turbine powered battle ships in WWII. It is considered necessary that longitudinal vibration be a subject of concept design. The main machinery items have long lead times, and any problems are to be uncovered early. Main propulsion system fore-and-aft natural frequencies tend to fall in the range of propeller blade rate excitation frequency. For short shafting systems, the one-node (first) mode can be easily coincident with blade rate excitation, but with the second mode well above. For long shaft systems, the system just cannot be designed with the first mode above the blade rate excitation frequency and must therefore be configured so that it lies far enough below. But then the second mode becomes of concern with long shaft systems.

The dominating uncertainly with regard to longitudinal vibration is the stiffness of the main thrust bearing and its "foundation." The thrust bearing "foundation" is the serial ship structure that deflects, as a spring, in response to the thrust transmitted through the thrust bearing. The thrust bearing on a diesel is normally located in the engine casing aft. The long engine casing provides some extra stiffness over that with the steam plant. It is the serial stiffness in the engine room and ship bottom structure that can be the critically weak link; recall that in serial stiffness addition the overall stiffness is less than the stiffness of the most flexible element. This supporting structure must be carefully designed early to properly place the first two system natural frequencies relative to blade rate excitation in order to avoid serious problems. The control is through shipyard responsibility for the design of the engine room bottom structure and machinery foundations.

The three-mass model of Fig. 4.2 can be used for first estimates of the first two shafting/ machinery system modes. Three masses are considered the minimum number needed for estimating the first two system modes with reasonable accuracy. Definition of the mass and stiffness data shown on the figure is as follows:

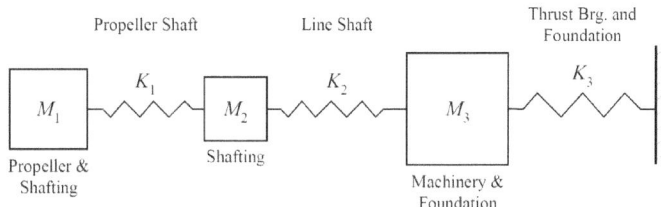

Fig. 4.2 3-mass longitudinal model of main propulsion system

M_1 lumped mass at the propeller in kg, composed of propeller mass, increased 60% for hydrodynamic added mass, one half of the propeller shaft weight

K_1 stiffness of propeller shaft in N/m, from propeller to coupling with line-shaft AE/ℓ_p

A shaft cross-sectional area in m^2

M_2 lumped mass at propeller shaft/line shaft coupling in kg, composed of one-half of propeller shaft mass plus on half of line-shaft mass. For very short propeller shafts, M_2 could be located in the line-shaft with mass and stiffness contributions appropriately adjusted

K_2 stiffness of line-shaft in N/m, from coupling to thrust bearing (thrust bearing assumed to be at aft end of engine casing)

M_3 lumped mass at thrust bearing in kg, composed of one-half of the line-shaft mass, the engine, including the thrust bearing, plus a thrust bearing/engine foundation structural weight allowance of 25%

K_3 stiffness of thrust bearing elements and engine foundation in N/m.

It is convenient to view the foundation stiffness K_3 as an unknown, to be determined so that the two natural frequencies lie at appropriate levels with respect to the blade-rate excitation frequency. The coupled equations of motion lead to the eigenvalue problem:

$$\begin{bmatrix} -\omega_n^2 M_1 + K_1 & -K_1 & 0 \\ -K_1 & -\omega_n^2 M_2 + K_1 + K_2 & -K_2 \\ 0 & -K_2 & -\omega_n^2 M_3 + K_2 + K_3 \end{bmatrix} \begin{vmatrix} \psi_{n1} \\ \psi_{n2} \\ \psi_{n3} \end{vmatrix} = \begin{vmatrix} 0 \\ 0 \\ 0 \end{vmatrix},$$

where ψ denotes the mode shape vector. It is necessary to expand the determinant of the coefficient matrix to form the characteristic equation whose roots are the three natural frequencies. First, define the following for convenience of notation:

$$\Omega_n = w_n^2, \ \Omega_{11} = \frac{K_1}{M_1}, \ \Omega_{12} = \frac{K_1}{M_1}, \ \Omega_{22} = \frac{K_2}{M_2}, \ \Omega_{23}\frac{K_2}{M_3}, \ \Omega_{33} = \frac{K_3}{M_3}$$

Then the characteristic equation from the determinant expansion is the following cubic in Ω_n:

$$\Omega_n^3 = \Omega_n^2(\Omega_{11} + \Omega_{12} + \Omega_{22} + \Omega_{23})$$
$$- \Omega_n[\Omega_{11}(\Omega_{22} + \Omega_{23}\Omega_{23}) + \Omega_{12}(\Omega_{23} + \Omega_{33}) + \Omega_{22}\Omega_{33}]$$
$$+ \Omega_{11}\Omega_{22}\Omega_{33} = 0$$

The unknown K_3, in Ω_{33}, can be calculated for specified distributions of Ω_n in the ranges of interest.

$$\Omega_{33} = \frac{\Omega_n[\Omega_n^2 + \Omega_n(\Omega_{11} + \Omega_{12} + \Omega_{22} + \Omega_{23}) - (\Omega_{11}\Omega_{22} + \Omega_{11}\Omega_{23} + \Omega_{21}\Omega_{23})]}{-\Omega_n^2 + \Omega_n(\Omega_{11} + \Omega_{12} + \Omega_{22}) - \Omega_{11}\Omega_{22}}$$

The corresponding value to the required foundation spring constant is:

$$K_3 = M_3\Omega_{33}$$

K_3 can then be plotted as a function of the arbitrary Ω_n to decide the stiffness of the structure that the shipyard is to design and build to provide the proper support stiffness for the system. It should be pointed out that on increasing the Ω_n from low values, K_3 will increase, and then will become negative as the second mode is reached. The stiffness subsequently turns back to positive again with further increasing frequency in the second mode. The relevant ranges are the ranges of positive K_3 only.

Example: Take the case used with the propeller example of the preceding sub-section, with the propeller RPM $= 120$ with 5-blades, implying a blade rate frequency of 10 Hz. The data used in this example are:

Propeller weight, $W_p = 24{,}098$ kg   Propeller shaft diameter, $d_{ps} = 9.45$ cm

Propeller shaft length, $L_{ps} = 15.24$ m   Line shaft diameter, $d_{ls} = 11.81$ cm

Line shaft length, $L_{ls} = 2134$ m   Engine weight, $W_e = 136{,}080$ kg

By the lumping scheme described above, the masses and springs are:

$M_1 = 65{,}770$ kg $M_1 = 6{,}182 \times 106$ N/m

$M_2 = 62{,}050$ kg $M_2 = 1{,}979 \times 106$ N/m

$M_3 = 204{,}940$ kg $M_3 = k_f$ to be determined

Figure 4.3 shows the plot of natural frequency versus thrust bearing and foundation stiffness for the first two modes.

The blade-rate exciting frequency on 4/3 is 10 Hz, so that the foundation stiffness is to be either sufficiently above or below approximately 15×10^8 N/m. The restricted frequency band is conventionally taken to be from 20% below the resonance to 20%

Fig. 4.3 Example of natural frequencies versus foundation stiffness

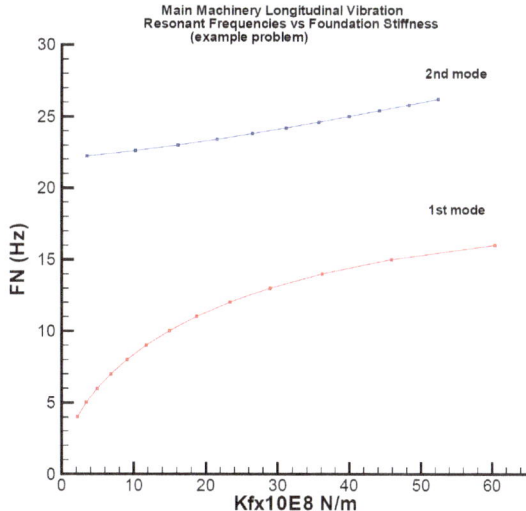

above. Therefore, by this criterion, the stiffness is to be either below 9.5×108 N/m or above 24×10^8 N/m. Recall that k_f is the serial stiffness of the thrust bearing and the foundation; the trust bearing stiffness is typically around 36×10^8 N/m. By the serial addition, this gives a higher required stiffness range for the foundation proper: below 12.9×10^8 or above 72×10^8 N/m. It is desirable to set the excitable natural frequencies above the excitation frequency range. Otherwise, a resonance falls within the operating range; in that case it must be assured that it is far enough below the full power blade-rate frequency.

In this particular example, which is not atypical, the second mode is not of relevance as it lies well above the exciting frequency range for all foundation stiffnesses. As for the first mode, experience has shown that it would be impractical for the shipbuilder to build a foundation with a stiffness as high as 72×10^8 N/m. The available choice would be a foundation stiffness no higher than 12.9×10^8 N/m.

Designing a thrust bearing foundation for a specified stiffness is no simple matter, with shear deflection of the girders and bending deflection of the inner bottom out to some distance away typically involved. The best recourse is probably to design the double bottom to be as deep and stiff as possible within other constraints. This then helps to limit the deflecting structure to the foundation proper above the double bottom where it can be dealt with more reliably in detailed structural design.

The following steps are therefore recommended in concept design:

(1) Approximate the constants in the 3-mass model and perform the calculation of the combined thrust bearing and foundation stiffness,
(2) Estimate the thrust bearing stiffness and perform the serial subtraction to establish a first estimate of the foundation stiffness. Graph the result in the form of Fig. 4.3.

(3) If PRPM and blade number N have been tentatively set, select a foundation stiffness from the graph for which neither of the natural frequencies are within 20% of the full power blade-rate excitation frequency,

(4) If PRPM and/or blade number have not been set, or (3) above cannot be achieved, select PRPM $\times$ N such that (3) is be achieved. Preferably, select PRPM and N such that the first mode resonant frequency falls at least 20% above the full power blade-rate excitation to avoid a critical in the power range. ($N = 5$ blades is the best choice for minimum blade-rate alternating thrust for conventional stern ships with conventionally configured propellers, as discussed in Chap. 3), and

(5) Provide the required foundation stiffness established to the hull structural designers for effecting in the preliminary/detailed structural design stages.

4.4   Superstructure Fore-and-AFT Vibration Excited

The movement of commercial ship engine rooms in the 1960's from amidship to the stern was a technological advancement in all respects except one: propeller-induced vibration. The movement was prompted by longitudinal strength as ships were becoming longer. But removing the structural discontinuity from the midship region and shortening the shafting system also reduced cost. However, wakes inherently became more irregular with the fuller sterns, and with the increasing ship power at that time, the role of propeller blade cavitation as a dominant vibration exciter quickly became apparent. In order to have unobstructed view over the ship bow from the aft-located bridge, elevation of the bridge atop a towering deckhouse was required. With this deckhouse necessarily mounted over the engine room cavity, adequate structural stiffness was difficult to incorporate. Then, with the house also located directly above the propeller, propeller-excited deckhouse vibration became the industries' troublesome problem.

As suggested in this chapter, Sect. 4.2, a hull girder beam analysis which ignores the dynamics of superstructure produces useful estimates of the hull girder lower natural frequencies for purposes of resonance avoidance with a main diesel engine. It is indeed fortunate that the lower rocking/bending natural frequencies associated with stern superstructures, which usually fall in the range of propeller blade-rate exciting frequencies, can, conversely, be estimated with useful accuracy by ignoring the dynamics of the hull girder. This is the case when the mass of the superstructure is small, relative to the effective mass of the hull girder near the coupled natural frequencies of interest. Any consideration of vibratory response, versus natural frequencies alone, must, on the other hand, allow for the dynamic coupling. This is based on consideration of the fact that, in the preponderance of cases, superstructure vibration is excited by the hull girder vibration at its base. The superstructure vibration mode of primary concern is a fore-and-aft rocking/bending mode excited most often through vertical blade-rate vibration of the hull girder. This mode can also be excited indirectly at blade-rate frequency by longitudinal excitation from longitudinal resonance of the shafting/machinery system via the main thrust bearing (in this chapter, Sect. 4.3).

For obtaining preliminary estimates of superstructure fore-and-aft rocking/bending natural frequencies the semi-empirical method of Hirowatari and Matsumoto (1969) has proved to have great utility (Sandstrom and Smith, 1979). This method was developed from correlation of simple analysis and measured fore-and-aft superstructure natural frequencies on approximately thirty ships. In this method, the fore-and-aft "fixed base" natural frequency of the superstructure (i.e., superstructure cantilevered from the main deck) is determined according to deckhouse type and height. The fixed base natural frequency is then reduced by a correction factor to account for the rotational flexibility of the underdeck supporting structure. Specifically, the procedure of Hirowatari is as follows:

(1) Select superstructure type from Fig. 4.4.
(2) Determine the superstructure height, h.
(3) Read f_∞(fixed-base natural frequency) as a function of h from Fig. 4.5.
(4) Read f_e/f_∞ (the correction factor) from Table 4.2.
(5) Compute f_e (the expected deckhouse natural frequency in the first fore-and-aft mode) from the following formula:

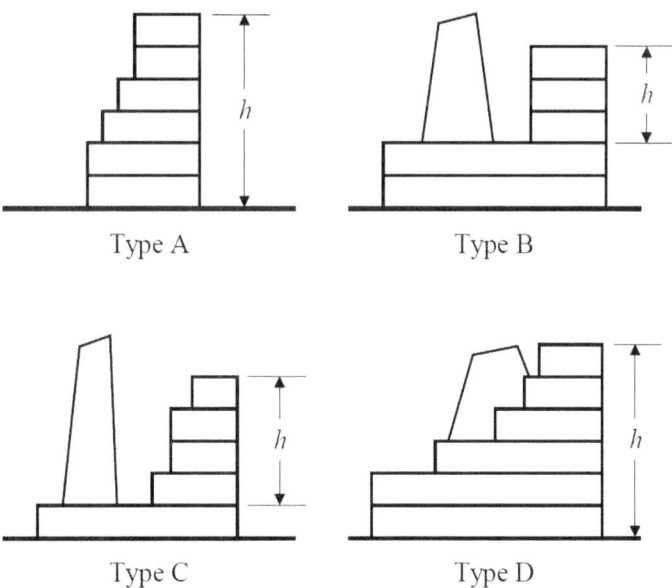

Fig. 4.4 Deckhouse types

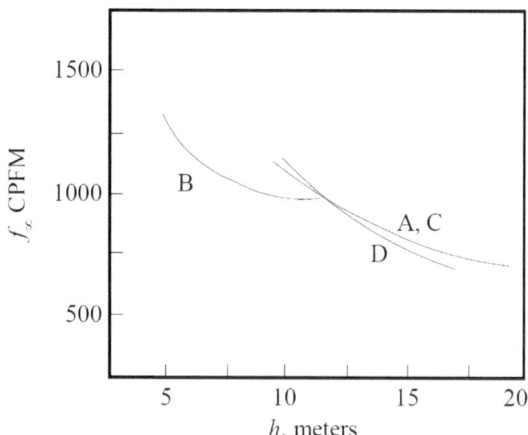

Fig. 4.5 Fixed-base superstructure natural frequencies

$$f_e = f_\infty (\frac{f_e}{f_\infty})$$

It is reported that this procedure generally produces results that are within 15 percent of measurements from shaker tests. However, the method becomes inapplicable when the superstructure type varies significantly from those given in Fig. 4.4. Furthermore, there is some uncertainty regarding the use of the correction factors for superstructure support flexibility given in Table 4.2, since the supporting structure may vary from deep beams to column supports to structural bulkheads. Despite these difficulties, the method seems to work quite well in most cases, considering the limited input that is required. This feature makes the Hirowatari Method particularly attractive in the concept design stage when the design data are sparse or non-existent.

The two basic effects influencing the fundamental fore-and-aft superstructure natural frequency are exemplified in this approach:

(1) Cantilever (fixed base) bending and shear of the superstructure as a beam over its height h, Fig. 4.5.
(2) Rocking of the superstructure as a rigid box on the effective torsional stiffness of its supporting structure.

Ordinarily, one of the superstructure main transverse bulkheads will be a continuation of one of the two engine room transverse bulkheads. The intersection of the continuous bulkhead and the deck identified with the superstructure base can usually be taken as the axis about which the rocking of the house occurs.

The fore-and-aft natural frequency of the superstructure due to the combined effects of rocking and bending/shear can be estimated using Dunkerley's equation (Thomson, 1973) as,

$$f_e \sqrt{\frac{1}{\frac{1}{f_\infty^2} + \frac{1}{f_R^2}}}$$

Here, f_∞ has been identified as the fixed base cantilever natural frequency, from Fig. 4.5, or by detail analysis. fR is the rocking natural frequency of the rigid superstructure, of height h, on its supporting stiffness,

$$f_R = 60/2\pi \sqrt{K_f/Jcpm}$$

$$fR = 60/2\pi \ Kf/Jcpm$$

J is the mass moment of inertia of the superstructure about the rocking axis and K_f is the effective torsional stiffness of the superstructure foundation, also about the axis of rotation. The Hirowatari procedure, in conjunction with the above formula, has utility in design or post design corrective studies where estimates must be made of the relative effects of structural changes. This is demonstrated by the following example:

Example: Assume that a conventional Type A superstructure, Fig. 4.4, has been concept designed and that for the selected engine and propeller at full power, PRPM = 100 and N = 5, giving a blade-rate excitation frequency of 500 cpm. The house height, h, is 15 m. Referring to Figs. 4.4 and 4.5 and Table 4.2:

$$f_\infty \cong 800 \ cpm$$

And

$$\frac{f_e}{f_\infty} = 0.625$$

This gives the estimated fore-and-aft house natural frequency, f_e, as:

$$f_e = 0.625 \ 800 = 500 \ cpm$$

which is precisely the full power blade rate excitation frequency. Proceed with the idea of stiffening the system in order to raise the natural frequency by the 20% minimum to 600 cpm. The rocking frequency is first estimated as,

$$f_R = \sqrt{\frac{1}{\frac{1}{f_e^2} - 1/f_\infty^2}} = 640 \; cpm$$

Then,

$$\frac{K_f}{J} = \frac{2\pi}{(60)^2 f_R^2} = 4490 \; rad^2/sec^2$$

Now assume that the mass of the house, m, has been estimated as 300 tonnes. Also assume that the house front is a continuation of the engine room forward transverse bulkhead, so that the house effectively rotates about its front lower edge, Fig. 4.6. Assume a radius of gyration, r, of the house about this axis of 10 m. The house mass moment of inertia, J, is then:

$$J = m_r^{-2} = 3 \times 10^7 \; kg - m^2$$

Fig. 4.6 Deckhouse stiffening

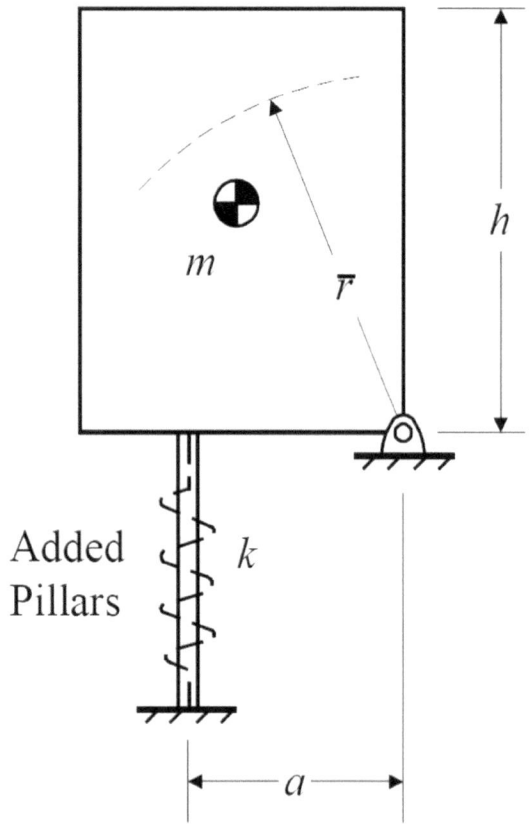

The effective rotational stiffness of the foundation is then estimated as:

$$K_f = 1.35 \times 10^{11} \ N - \frac{m}{rad}$$

Proceeding, assume that stiffening is proposed in the form of two parallel pillars made up of 20 cm extra heavy steel pipe, each 6 m long, and located under the house side bulkheads, as indicated on Fig. 4.6.

The effective axial stiffness of the parallel pillars is calculated to be $k = 5 \times 10^8$ N/m. The pillars are located at a distance $a = 5$ m aft of the forward bulkhead, so that the incremental rotational stiffness added by the pillars is:

$$\delta K_f = ka^2 = 1.25 \times 10^{10} \ N - \frac{m}{rad}$$

The stiffness of the stiffened foundation is:

$$K_{f\prime} = K_f + \delta K_f = 1.475 \times 10^{11} \ N - \frac{m}{rad}$$

which represents a 9.3 percent increase. The new rocking frequency becomes:

$$f_{R\prime} = f_R \sqrt{1.093} = 669 \ cpm$$

Then, the house fore-and-aft natural frequency is raised to:

$$f_{e\prime} = \sqrt{\frac{1}{\frac{1}{f_\infty^2} + 1/f_R^2}}$$

This represents a 2.6 percent increase over the value of 500 cpm without the pillars, and far less than the 600 cpm judged to be needed. In order to achieve $f_{e\prime} = 600$ cpm, reversing this last calculation gives $f_R = 907$ cpm, and $K_f = 2.\ 72 \times 10^{11}$ N-m/rad. This is about double the rotational stiffness of the original design. It would be essentially impossible to double the support structural stiffness if standard design features were employed.

In conclusion, the simple analysis of this example demonstrates the common misconception that vibration problems are always eliminated by stiffening the structure. This may be true for light structures such as mast, hand-rails, and the like, but with major substructures, such as a deckhouse, once the scantlings and principal dimensions are set incorrectly, the problem tends not to be correctable by simple stiffening, as the preceding example suggests. It may be possible to reduce stiffness to avoid a resonance, which was the spirit of the previous example on thrust bearing foundation stiffness in this chapter, Sect. 4.3. But reductions in stiffness are generally undesirable. Stiffness reduction can result in excessive static deflection. As in the thrust bearing foundation stiffness example, the simplest recourse might be to change the engine RPM and/or the propeller blade number. In this current example, a switch from a five to a 6-bladed propeller with no structural changes would place the resonance in the operating range, but at the desired 20% below the new full power blade-rate exciting frequency of 600 cpm.

The following specific steps are suggested for reducing the potential danger of excessive propeller induced deckhouse vibration at the concept design level.

(1) Perform the preceding analysis on the candidate deckhouse and support structure design to predict the fore-and-aft rocking bending/shearing natural frequency,
(2) Compare the natural frequency with the blade-rate exciting frequency of any proposed engine and propeller characteristics. If the engine and/or propeller have not been proposed, note the blade rate frequency to be avoided in that ultimate selection, and
(3) If the engine and propeller characteristics have been proposed and there is reluctance to change due to conflicts with other design considerations, establish the stiffness range of the deckhouse supporting structure calculated as necessary to avoid resonance within 20% of the full power blade rate frequency. This is just as recommended with the longitudinal vibration issue at the concept design stage. This defers the ultimate implementation to the preliminary/detailed designers who will at least have been made aware of the potential difficulty going in.

5.1 Introduction

5.1.1 Scope and Objective

The design and construction of a ship free of excessive vibration continues to be a major concern and, as such, it is prudent to investigate, through analysis, the likelihood of vibration problems early in the design stage. Vibration analysis is aimed at the confirmation of the many design considerations associated with:

- Stern configuration,
- Main propulsion machinery,
- Propeller and shafting system, and
- Location and configuration of major structural assemblies.

The ship hull structure includes the outer shell plating and all internal members, which collectively provide the necessary strength to satisfactorily perform the design functions in the expected sea environment. The hull structure responds as a free-free beam (both ends free) when subjected to dynamic loads. The vibration induced by the propulsion system is a common source of ship vibration. The vibration from this source manifests itself in several ways. Dynamic forces from the shafting system are transmitted to the hull through shaft bearings. The propeller induces fluctuating pressures on the surface of the hull, which induces vibration in the hull structure. The main and auxiliary engines can directly cause vibrations through dynamic forces transmitted through their supports and foundations. The response to this forcing can cause the vibration of the hull girder, deckhouse, deck and other structures, local structures and equipment. When attempting to determine the source of vibration, it is necessary to establish the frequency of excitation

© The Author(s), under exclusive license to Springer Nature Switzerland AG 2025   43
F. Karkori, *Ship Vibration 1*, Synthesis Lectures on Ocean Systems Engineering,
https://doi.org/10.1007/978-3-031-75072-4_5

and to relate the frequency of excitation to the shaft rotational frequency by determining the number of oscillations per shaft revolution. The main engine-induced unbalanced excitations encountered with slow-speed diesel-driven ships are the primary and secondary free engine forces and moments. The engine manufacturer is to provide the magnitude of these forces and moments.

The response of the hull structure may be resonant or non-resonant. The hull structure will normally vibrate in the following modes:

- Vertical bending,
- Horizontal bending,
- Torsional (twist),
- Longitudinal, and
- Coupling exists between horizontal and torsional modes, especially in containerships.

Typical major substructures include deckhouses, main deck structures and large propulsion machinery system, etc., which because of the direct coupling with hull structure vibration can significantly influence the total or global pattern of ship vibration. In analysing vibration patterns of such large complex structures, it is necessary to identify the principal reason for observed excessive vibration. Excessive vibration of a major substructure may be the result of structural resonance in the substructure or in the attachment detail for the substructure and hull structure.

Local structural components are the minor structural assemblies, relative to major substructures previously referred to. Local structural component may be identified as decks, bulkheads, platforms, handrails, minor equipment foundations, or main engine coupled with its seating and inner bottom structure, etc., and are components of larger structures (major substructures) or of the entire hull structure. Most ship vibration problems occur in local structural components and are the result of either strong inputs received from the parent structure amplified by resonance effects in the local structure or are the response to vibratory forces generated by mechanical equipment attached to the local structure.

Local structural panels are basic elements forming local structural components. Local structural panels are identified as plate panels between stiffeners, stiffener panels enclosed by strong supports, or girder panels enclosed by bulkheads and/or decks. Typically, the natural frequencies of local structural components or local structural panels are to be out of the range between 85 and 115% of major excitation frequencies to avoid resonance. Thus, local structural vibration levels will not be major contributors to vibration levels of critical areas.

This section describes the analysis procedure for propeller and main engine induced ship hull vibration using a three-dimensional finite element method. The objective of vibration analysis is to determine the overall vibration characteristics of the hull structure and major superstructures so that areas sensitive to vibratory forces may be identified and assessed relative to standard acceptance criteria on vibration level.

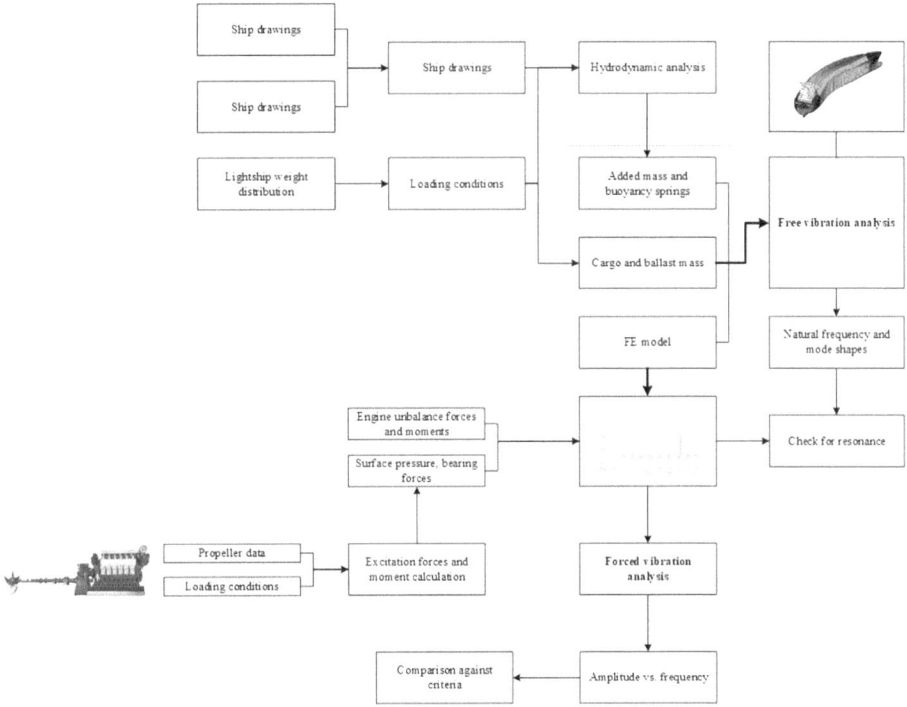

Fig. 5.1   Procedure to perform ship vibration analysis

5.1.2   Procedure Outline of Ship Vibration Analysis

The flowchart of the procedure is shown in Fig. 5.1. The procedure recommended is to perform the vibration analysis using a three-dimensional finite element model representing the entire ship including the deckhouse and main propulsion machinery system. Both free and forced vibration analyses are included.

5.2   Finite Element Modelling

5.2.1   Global Model

5.2.1.1   General
Ship structures are complex and may be analysed after idealisation of the structure. Several simplifying assumptions are made in the finite element idealisation of the hull structure. The modelling requirements are that all significant structural sections are to be captured and deflection/velocity/acceleration are to be sufficiently predicted.

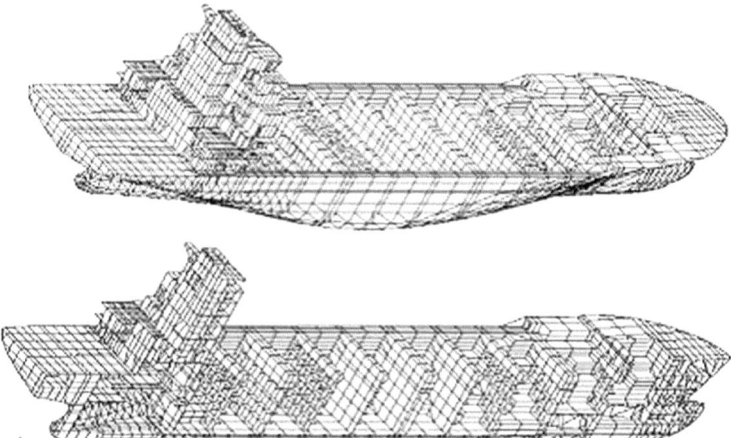

Fig. 5.2 Global FE model example

A three-dimensional finite element model representing the entire ship hull, including the deckhouse and machinery propulsion system, needs to be developed for vibration analysis. If a global model exists from any previous tasks such as stress analysis, it needs to be conditioned for vibration analysis.

5.2.1.2 Mesh Size

A judicious selection of grid points, elements and degrees of freedom may be used to fully represent the elastic and inertia properties of the structure while keeping the complexity of the data generation and the overall size of the model within manageable limits. Typically, a mesh size of three to four stiffener spacing for the overall ship structure is acceptable except for the stern and deckhouse structure. Bending plates and beam elements are to be used to achieve better structural stiffness representation. A finer mesh may be used in the areas of interest. Typical examples of such finite element models can be seen in Fig. 5.2.

5.2.2   Engine, Propeller Shaft and Stern/Skeg

5.2.2.1 Main Engine

Care is to be taken when modelling the engine and surrounding structure. The engine foundation is to be modelled in detail. The engine can be modelled using solid or shell elements such that the mass, C.G and the engine stiffness matches the specifications of the engine. By assigning different densities of the material, mass and centre of gravity of the engine can be controlled. In order to match the engine stiffness, different Young's moduli can be used for longitudinal and transverse structural members, respectively, based on the engine stiffness data provided by the engine manufacturer. An example of simplified diesel engine FE model is shown in Fig. 5.3. An example of a turbine engine FE model is shown in Fig. 5.4. Main engine top stays can be modelled as spring elements with spring coefficients obtained from manufacturer.

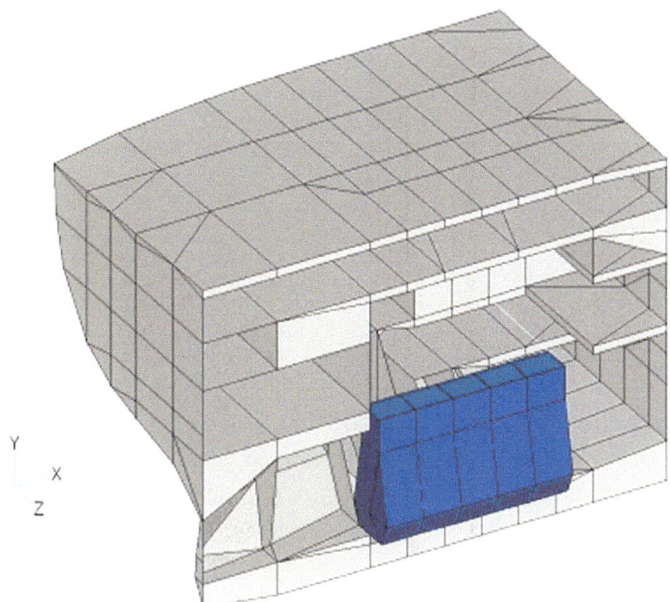

Fig. 5.3 Engine model example

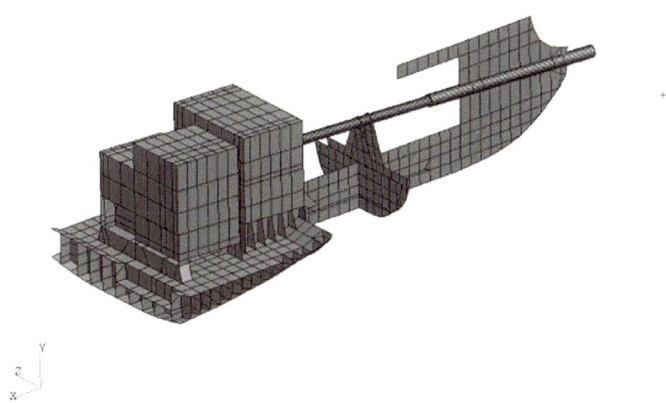

Fig. 5.4 Turbine engine and propeller shaft modelling example

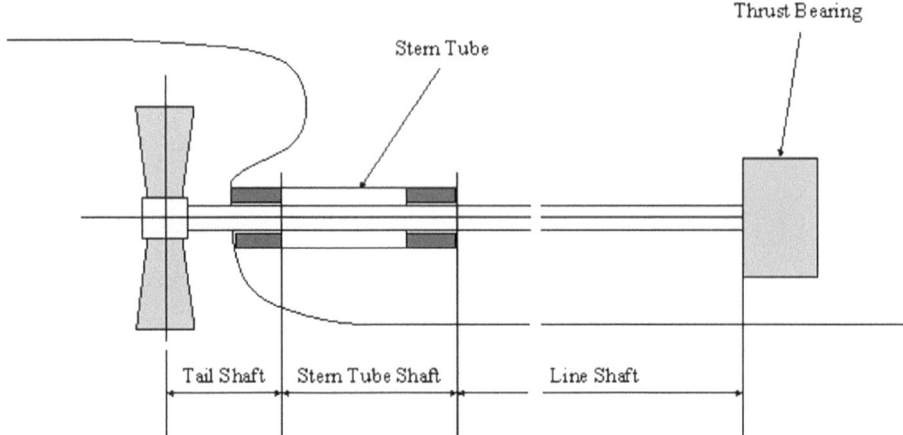

Fig. 5.5 Propeller shaft

5.2.2.2 Propeller Shaft and Propeller
The propeller shaft is to be explicitly modelled such that it represents the proper stiffness of the shaft. It may be modelled with beam or bar elements. Care is to be taken when modelling the support bearings. The transverse supporting structure is to be modelled with the help of membrane shell elements. The tail shaft length is to be from the centre of the propeller hub as shown in Fig. 5.5. The propeller may be modelled as a mass element.

5.2.2.3 Stern/Skeg Structure
Proper care is to be taken when modelling the skeg structure. Typically, the skeg structure is a thick casting. If it is modelled with shell elements, the thickness is to be increased appropriately to account for the stiffness of that structure.

5.2.3 Lightship Weight Distribution

Lightship weight distribution is an important factor in any vibration analysis. Results can change dramatically if it is not properly represented. Typically, masses of all the heavy equipment are to be modelled as mass elements and are to be placed at well-supported nodes such that the total mass and centre of gravity of that equipment are maintained. This distribution is to be as accurate as possible in the areas of interest. In general, the CG from the FE model and the CG from the T & S booklet is to be within 0.5% of the total ship length.

The deckhouse superstructure weight and mass distribution, longitudinally and vertically, must also be accurately represented. Typically, the non-structural masses that account for furniture, outfitting, and deck coverings in deckhouse superstructure are to be modelled by changing local structural density or modelled as mass elements uniformly placed at well-supported nodes. This distribution is to be provided by the ship designer or shipbuilder. If an accurate distribution is not available, sensitivity studies are to be conducted to determine the most critical values between 40 kg/m^2 to 100 kg/m^2.

5.2.4   Cargo, Water Ballast in Tanks and Fuel Oil in Tanks

Mass elements can be applied to the tank boundaries to represent cargo, water ballast in tanks, and fuel oil in tanks.

5.2.5   Local Structural Component Models

For local structural components, the FE models should also include surrounding structures which are to be extended to the adjacent strong supports, such as primary transverses or girders, bulkheads, or decks, etc.

For example, for local vibration analysis of the navigation deck, the FE model should include not only the navigation deck but also the surrounding bulkheads between the navigation deck and adjacent decks. In the calculations, fixed boundary conditions are applied to the FE model in order to reduce the modes of the extended structures. The mesh size can range from one or two elements between stiffeners to about six elements between stiffeners.

5.2.6   Local Structural Panel Models

A number of local structural panels are selected from the various decks, platforms, flats and bulkheads within the superstructure and regions of the hull for the natural frequency calculations. The panels selected are either typical locations or those locations expected to have the lowest natural frequencies in the areas of concern. In the calculations, pinned boundary conditions are applied to stiffener panels and plate panels. For girder panels, suitable boundary conditions should be assessed from the local design arrangements if pinned boundary conditions result in lower natural frequencies than major excitation frequencies. The mesh size can range from three elements between stiffeners to about six elements between stiffeners.

5.3   Loading Condition

5.3.1   Selection of Loading Conditions and Ship Speed

The objective of the vibration analysis is to investigate the ship vibration performance at intended service conditions. Therefore, the loading conditions, such as Full Load Condition and Ballast condition, in which the ship operates at ship design speed, will be the focus of the vibration analysis. In addition, it is often desirable to investigate the sea trial condition for the purpose of calibrating calculated numerical results with measurements. Typically, considering the analysis efforts to be taken, it is recommended that vibration analysis be performed for two selective conditions, either.

- Full load condition and sea trial condition, or
- Full load condition and ballast load condition.

Depending on the loading condition, the mass of the cargo and ballast is then distributed in the structural model using mass elements. The corresponding added mass and the buoyancy springs are then calculated and added to the model.

5.3.2   Added Mass

For local vibration analysis, when local structures are in contact with fluid, the added mass effect of the fluid must be considered in the calculation. The added mass can be considered either through virtual mass method using boundary element method (BEM) such as 'MFLUID' card of MSC NASTRAN or using empirical formulas. For global vibration analysis, a three-dimensional seakeeping analysis program may be used. The analysis includes three main tasks:

- *Development of a hydrodynamic panel model:* The hydrodynamic panel model is to represent the geometry of the ship's hull below the still-water line when using a linear seakeeping program. It is recommended that a total of about 2000 panels be used for the ship's hull surface, including port and starboard sides. A computer program may generate the panel model using the ship's offsets. The main particulars required for the hydrodynamic model includes ship displacement, drafts, location of centre of gravity and radii of gyrations,
- *Hydrodynamic analysis:* The purpose of the analysis is to obtain the distributed added mass. In general, the natural frequency of the ship is much higher than the wave frequencies considered in the seakeeping analysis. To obtain the added mass at such high frequencies, infinite frequency added mass option is to be used, and

- *Mapping hydrodynamic results onto FE model:* The distributed added mass from the seakeeping analysis is represented as an added mass on each hydrodynamic panel model. An interface program can be used to map the heave added mass onto the FE model. The user needs to check that the total added mass on the FE model is equal to the total added mass on the hydrodynamic panel model.

Alternative approaches may be considered acceptable on a case-by-case basis.

5.3.3 Buoyancy Springs

The effect of buoyancy on hull vibration is generally regarded as small. Where it is necessary to consider the buoyancy effect, it may be modelled by adding rod elements to the wetted surface of the model. The rod elements work as springs and the total stiffness of the rod elements is to be equivalent to the ship's vertical buoyancy stiffness.

5.3.4 Special Conditions

Shipyards and/or ship owners may consider additional loading conditions to be investigated during the vibration analysis process.

5.4 Free Vibration

5.4.1 Analysis Procedure

Computation of the natural frequencies and mode shapes is to be performed by solving an eigenvalue problem. The natural frequencies (eigenvalues) and corresponding mode shapes (eigenvectors) of the three-dimensional finite element model can be obtained by solving the following equation of motion:

$$[M]\{\ddot{u}(t)\} + [C]\{\dot{u}(t)\} + [K]\{u\{(t)\} = \{F(t)\}$$

where

C	damping matrix
$\ddot{u}$	column matrix of accelerations
$\dot{u}$	column matrix of velocities
u	column matrix of displacements
F	column matrix of harmonic forces.

For free vibration, damping $[C]$ and forces $\{F\}$ are zero. The solution would then be from:

$$[K]\{\Phi\} = \omega^2[M]\{\Phi\}$$

where

K   symmetrical stiffness matrix
M   diagonal mass matrix
Φ   column mode shape matrix
ω   natural frequency

This problem can be solved by normal mode analysis. An important characteristic of normal modes is that the scaling or magnitude of the eigenvectors is arbitrary. Mode shapes are fundamental characteristic shapes of the structure and are therefore relative quantities. Examples of mode shapes of a typical LNG carrier are shown in Figs. 5.6 and 5.7. The natural frequencies obtained from the analysis can then be compared to the excitation frequencies to check for resonance.

5.4.2 Checking Points

5.4.2.1 Mode Shapes
When performing free vibration analysis, it is important to validate the calculated results by investigating their mode shapes. There are several checking points:

- The first six mode shapes are to be rigid body modes and these six rigid body mode shapes, namely Pitch, Roll, Yaw, Surge, Heave and Sway, are not to display elastic distortion. Frequency is generally very low, well below the first elastic natural mode. Any mixing of rigid body modes and/or missing rigid body mode(s) would be a good indication of an erroneous FE modelling, especially when incorrect multi-point constraints are applied to the FE model.
- The first several elastic modes are to be for the hull structure. For typical commercial ships, there is usually no local modes below 4 Hz. Any "local modes" with low natural frequencies would be an indication of incorrect FE modelling. Too low frequencies are usually an indication of too high a mass or too low stiffness, or both.

It is a usual practice to investigate mode shapes up to at least twice the blade-rate frequency.

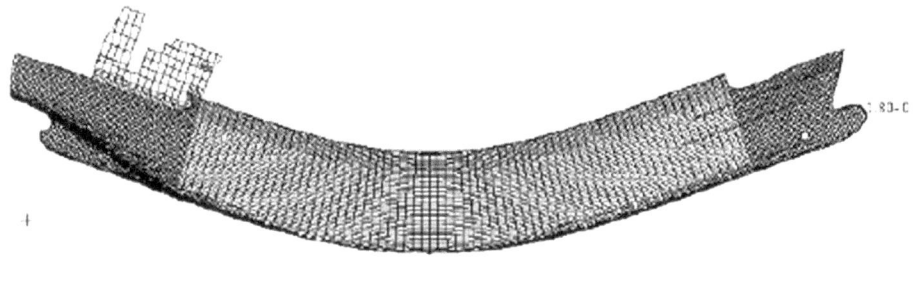

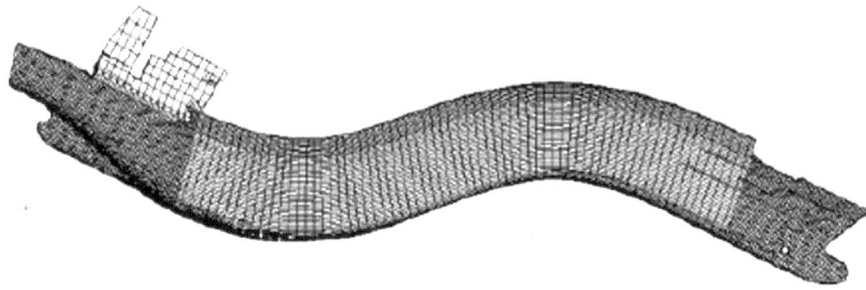

Fig. 5.6   First two vertical mode shapes

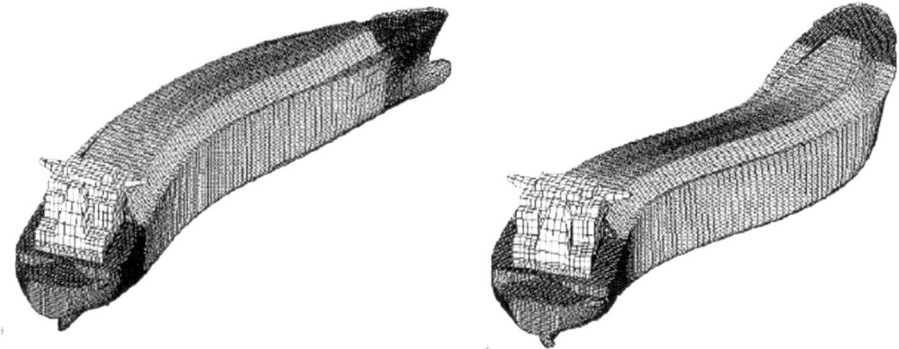

Fig. 5.7   First two horizontal mode shapes

5.4.2.2   Natural Frequencies

Another checking point is to check the calculated natural frequencies from the free vibration analysis. Kumai's formula in Chap. 4, Sect. 5.3 for two-node vertical vibration may
be used to preliminarily evaluate the calculated two-node natural frequency in order to
verify that the FEM frequencies are in a proper range.

5.5   Propeller Excitation

5.5.1   Introduction

Excitation forces from the propeller are transmitted into the ship via the shaft line and also
in the form of pressure pulses acting on the ship hull surface above the propeller. Whereas
propeller shaft forces (bearing forces) are the most significant factor for vibrations of shaft
lines, the predominant factor for vibrations of ship structures are the pressure fluctuations
on the hull surface (hull surface forces).

5.5.2   Propeller Shaft Forces

5.5.2.1   General

Propeller shaft forces (bearing forces) are fluctuating load effects caused by the non-
uniform wake inflow to the propeller. Usually, bearing force fluctuations transmitted from
propeller to propulsion shafting are slightly affected by cavitation processes, if blade
cavitation is not severe. In the absence of high shaft inclination, the magnitude of the
bearing forces depends on the characteristics of the wake field, the geometric form of
the propeller (in particular the skew and blade number), the ship speed and the rotational
speed of the propeller. For a given application, the forces and moments generated by a
fixed pitch propeller without severe cavitation are, in general, proportional to the square of
the revolutions, since for a considerable part of the upper operational speed range the ship
will work at a nominally constant advance coefficient. Generally, the thrust fluctuation
can be as high as 10% of the mean thrust, but it is usually between 2 to 8%. The force
fluctuations in transverse and vertical directions are between about 1 to 2% of the mean
thrust. In most cases, the moment fluctuation about the transverse axis is predominant
($5 \sim 20\%$ of the mean torque) compared to fluctuations of the moment about the vertical
axis and of the torque ($1 \sim 10\%$).

5.5.2.2   Previous Studies

In the 1980's, an analytical investigation of first and second blade rate bearing forces/
moments was performed for 20 ships (Veritec, 1985). The results of these calculations
are useful for making preliminary estimates of the dynamic forces at the early stages of

design. Table 5.1 shows the results of these calculations in terms of the mean values and their ranges. In the Table, each of the six loading components is expressed in terms of the mean thrust T_0 or the mean torque Q_0.

Propeller bearing forces and moments can excite various modes of vibration. For computations of axial and torsional vibrations of the shaft line, fluctuations in thrust and torque are to be taken into account. Bending vibrations of the shaft are influenced by transverse forces in horizontal and vertical directions as well as by bending moments about the corresponding axes.

Table 5.1 Propeller bearing forces and moments for 20 real ship case study

				Blade number		
				4	5	6
Blade rate frequency component	Thrust	$F_{x(1)}$	Mean	$0.084\ T_0$	$0.020\ T_0$	$0.036\ T_0$
			Range	$\pm0.031\ T_0$	$\pm0.006\ T_0$	$\pm0.002\ T_0$
	Vertical force	$F_{z(1)}$	Mean	$0.008\ T_0$	$0.011\ T_0$	$0.003\ T_0$
			Range	$\pm0.004\ T_0$	$\pm0.009\ T_0$	$\pm0.002\ T_0$
	Horizontal force	$F_{y(1)}$	Mean	$0.012\ T_0$	$0.021\ T_0$	$0.009\ T_0$
			Range	$\pm0.011\ T_0$	$\pm0.016\ T_0$	$\pm0.004\ T_0$
	Torque	$M_{x(1)}$	Mean	$0.062\ Q_0$	$0.011\ Q_0$	$0.030\ Q_0$
			Range	$\pm0.025\ Q_0$	$\pm0.0008\ Q_0$	$\pm0.020\ Q_0$
	Vertical moment	$M_{z(1)}$	Mean	$0.075\ Q_0$	$0.039\ Q_0$	$0.040\ Q_0$
			Range	$\pm0.050\ Q_0$	$\pm0.026\ Q_0$	$\pm0.015\ Q_0$
	Horizontal moment	$M_{y(1)}$	Mean	$0.138\ Q_0$	$0.125\ Q_0$	$0.073\ Q_0$
			Range	$\pm0.090\ Q_0$	$\pm0.085\ Q_0$	$\pm0.062\ Q_0$
Twice blade rate frequency component	Thrust	$F_{x(2)}$	Mean	$0.022\ T_0$	$0.017\ T_0$	$0.015\ T_0$
			Range	$\pm0.004\ T_0$	$\pm0.003\ T_0$	$\pm0.002\ T_0$
	Vertical force	$F_{z(2)}$	Mean	$0.008\ T_0$	$0.002\ T_0$	$0.001\ T_0$
			Range	$\pm0.004\ T_0$	$+0.002\ T_0$	$\pm0.001\ T_0$
	Horizontal force	$F_{y(2)}$	Mean	$0.001\ T_0$	$0.006\ T_0$	$0.003\ T_0$
			Range	$\pm0.001\ T_0$	$\pm0.003\ T_0$	$\pm0.001\ T_0$
	Torque	$M_{x(2)}$	Mean	$0.016\ Q_0$	$0.0014\ Q_0$	$0.010\ Q_0$
			Range	$\pm0.010\ Q_0$	$\pm0.008\ Q_0$	$\pm0.002\ Q_0$
	Vertical moment	$M_{z(2)}$	Mean	$0.019\ Q_0$	$0.012\ Q_0$	$0.007\ Q_0$
			Range	$\pm0.013\ Q_0$	$\pm0.011\ Q_0$	$\pm0.002\ Q_0$
	Horizontal moment	$M_{y(2)}$	Mean	$0.040\ Q_0$	$0.080\ Q_0$	$0.015\ Q_0$
			Range	$\pm0.036\ Q_0$	$\pm0.040\ Q_0$	$\pm0.002\ Q_0$

5.5.3 Hull Surface Forces Induced by Propeller Cavitation

5.5.3.1 General

Propeller-induced hull surface pressure fluctuations are more significant than shaft bearing forces. In merchant ships, for which a certain degree of propeller cavitation is generally tolerated for the sake of optimising the propeller efficiency, about 10% of propeller-induced vibration velocities are caused by bearing forces, whereas approximately 90% are due to pressure fluctuations, or hull surface forces. In the design of ships having propellers with weak cavitation, the ratio may be reversed, while at the same time the absolute excitation level is much lower. Pressure fluctuation is generally proportional to the acceleration of the propeller cavity volume. Calculation of the cavity volume may require knowledge of the pressure distribution on the propeller blade, both in the radial and circumferential direction. The flow condition in the tip region of the propeller blades has a particularly strong influence on cavitation processes. The formation and subsequent detaching of tip vortices additionally complicate the flow condition at the blade tips. Computer programs for the prediction of cavitation volumes are correspondingly complex.

To avoid strong cavitation-induced pressure amplitudes, the volume curve is to exhibit the smallest possible curvatures (note that pressure fluctuation is proportional to the acceleration of cavity volume). This can be achieved by influencing the wake (minimising wake peaks) and by a suitable choice of propeller geometry. However, improved cavitation characteristics are usually offset for by reductions in efficiency. Selection of a larger area-ratio and reduction in propeller tip loading by selection of smaller pitch and camber at the outer radii are the most effective measures.

In addition, by means of skew, a situation can be achieved where the individual profile sections of a propeller blade are not all subjected to their maximum loading at the same time, but instead the volume curve is rendered uniform by the offset in the circumferential direction. Some concepts tolerate comparatively severe cavitation and are aimed at making the growth and collapse of the cavitation layer as slow as possible.

From experience, the pressure amplitude above the propeller alone is not adequate to characterise the excitation behaviour of a propeller. Therefore, no generally valid limits can be stated for pressure fluctuation amplitudes. These amplitudes depend not only on technical constraints such as achievable tip clearance of the propeller, power to be transmitted, etc., but also on the geometry-dependent compromise between efficiency and pressure fluctuation. Nevertheless, pressure amplitudes at blade frequency of 1 to 2, 2 to 8 and over 8 kPa at a point directly above the propeller can be categorised as low, medium, and high, respectively. Total vertical force fluctuations at blade frequency, integrated from pressure fluctuations, would range from 10 kN for a small ship to 1,000 kN for a high-performance container ship. Whether these considerable excitation forces result in high vibrations depends on the dynamic characteristics of the ship's structure and can only be judged on the basis of a forced vibration analysis.

5.5.3.2 Empirical Formulae

There are three methods for predicting hull surface pressure: by empirical methods, by calculations using advanced theoretical methods and by experimental measurements. With regard to empirical methods, the most well-known and adaptable method is that of Holden, et al. (1980). This method is based on the analysis of full-scale measurements for some 72 ships. The method is intended as a first estimate of the likely hull surface pressures using a conventional propeller form. Regression based formula for estimation of the non-cavitating and cavitating pressure are proposed as follows by Holden, et al.

$$p_o = \frac{(ND)^2}{70} \frac{1}{Z^{1.5}} \left(\frac{K_o}{\frac{d}{R}} \right) \ N/m^2 \quad \text{(non-cavitating pressure)}$$

and

$$p_c = \frac{(ND)^2 V_s (w_{T\max} - w_e)}{\sqrt{(h_a + 10.4)}} \left(\frac{K_c}{\frac{d}{R}} \right) \ N/m^2 \ \text{(cavitating pressure)}$$

where

N	propeller rpm
D	propeller diameter, in m
Vs	ship speed, in m/s
Z	blade number
R	propeller radius, in m
d	distance from 0.9R to a position on the submerged hull when the blade is at the TDC (Top Dead Centre) position (m)
w_{Tmax}	maximum value of Taylor wake fraction in the propeller disc
w_e	mean effective full scale Taylor wake fraction
h_a	depth of shaft centreline
K_o	$1.8 + 0.4d/R$ for $d/R \leq 2$
K_c	$1.7 + 0.7d/R$ for $d/R < 1$
K_c	1.0 for $d/R > 1$

The total pressure impulse, which combines both the cavitating 'p_c' and non-cavitating 'p_o' components, is then calculated from:

$$p_z = \sqrt{(p_o^2 + p_c^2)}$$

Empirical methods of this type are particularly useful as a guide to the expected pressures. They should not, however, be regarded as a definitive solution, because differences, sometimes quite substantial, will occur when correlated with full-scale measurement. For example, pc and po regression formulae give results having a standard deviation of the order of 30% compared to the base measurements from which it was derived.

5.5.3.3 Theoretical Method

In the case of a more rigorous calculation, more details can be taken into account, which results in a higher level of accuracy. Theoretical models, which would be used in association with this form of analysis, can be broadly grouped into the lifting surface or vortex lattice categories. In particular, unsteady lifting surface theory is a basis for many advanced theoretical approaches in this field. Notwithstanding the ability of analytical methods to provide a solution, care must be exercised in the interpretation of the results since these are particularly influenced by factors such as wake scaling procedures, the description of the propeller model and the hull surface, the distribution of solid boundary factors and the harmonic order of the pressure considered in the analysis. Furthermore, propeller calculation procedures assume a rigid body condition for the hull, and as a consequence do not account for the self-induced pressure resulting from hull vibration.

These have to be taken into account by other means; typically, finite element models of the hull structure. As a consequence of all of these factors considerable care is to be exercised in interpreting the results. Analysis procedures for direct calculations of propeller-induced loads are briefly described in this chapter, Sect. 5.4. For more details, refer to the ABS RPD2005-07 Technical Report '*An Integrated Computational Process for Cavitating Propeller Induced Loads*'.

5.5.3.4 Model Tests

Model tests to predict hull surface pressures can be conducted in either cavitation tunnels or specialised facilities such as depressurised towing tanks. In the past, the arrangement in a cavitation tunnel comprised a simple modelling of the hull surface by a flat or angled plate above a scale model of the propeller. Although this technique is still used in some establishments, a more enlightened practice is to use a partial model with shortened body. Wake screens are typically used to fine-tune the wake field when a partial model is used for wake measurements. In some large facilities, however, a full-length ship model may be used. The advantage of using a model of the actual hull form is twofold: firstly, it assists in modelling the flow of water around the hull surface and does not require wake screens, and secondly it makes the interpretation of the measured hull surface pressure easier since the real hull form is simulated.

In order to interpret model test results, reference can be made to dimensional analysis, from which it can be shown that the pressure at a point on the hull surfaces above a propeller has a dependence on the following set of dimensional parameters:

$$p = \rho n^2 D^2 \Phi \left(\frac{J, K_T, \sigma, R_n, F_n, z}{D} \right)$$

where

ρ water density in ton/m^3

n propeller rps

D propeller diameter in m

J advance coefficient

K_T propeller thrust coefficient

σ cavitation number

R_n Reynolds number

F_n Froude number

z distance from propeller to point on the hull surface in m

As a consequence of the above equation, a pressure coefficient K_p can be defined as follows:

$$K_p = \frac{p}{\rho n^2 D^2}$$

As defined, dimensionless hull surface pressure is a function of propeller loading, cavitation number, geometric scaling, Froude number and Reynolds number.

In propeller cavitation tests, the test conditions are usually chosen such that the average propeller thrust loading (expressed by K_T and J-identity) is equal for model and full scale. In addition, the pressure is lowered to such a level that model and full-size cavitation number are equal at corresponding points in the propeller disc. Furthermore, if the model test needs to fulfil Froude number identity, the test can only be realised in a depressurised towing tank or a tunnel with free surface. For a cavitation tunnel without a free surface, a rate of rotation for model scale is chosen within practical limits related to the tunnel capacity, the particular test set-up and the ranges of static pressure to be adjusted.

Although cavitation number and Froude number can be satisfied in most model tests, Reynolds number identity can seldom be fulfilled. This causes the well-known scale effect on the model test measurements. To take the scale effect into account in the model measurement, sometimes leading-edge roughness for model propeller is used for tripping the flow over the propeller blades to turbulence. However, it also should be noted that scale effects exist not only on the flow around the propeller blade surface but also on the inflow toward propeller. Figure 5.8 shows the comparisons of maximum propeller induced pressure between mode scale and full-scale data. As seen, the discrepancies are notable. Usually, in order to obtain reliable model test results, model test measurement is to be appropriately calibrated based on full-scale data.

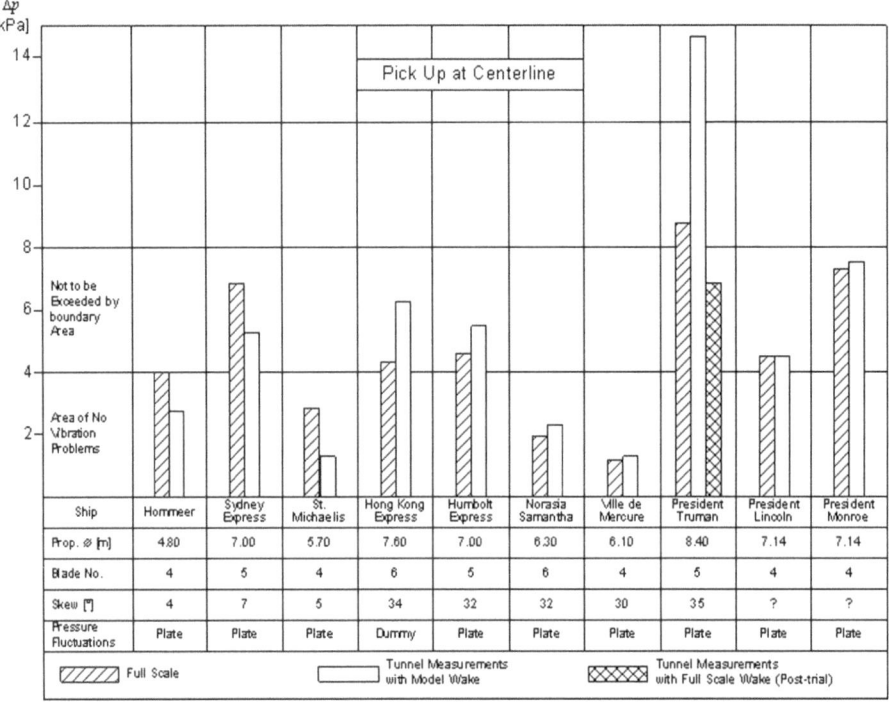

The chart table contains the following data:

Ship	Hommeer	Sydney Express	St. Michaelis	Hong Kong Express	Humbolt Express	Norasia Samantha	Ville de Mercure	President Truman	President Lincoln	President Monroe
Prop. ⌀ [m]	4.80	7.00	5.70	7.60	7.00	6.30	6.10	8.40	7.14	7.14
Blade No.	4	5	4	6	5	6	4	5	4	4
Skew [°]	4	7	5	34	32	32	30	35	?	?
Pressure Fluctuations	Plate	Plate	Plate	Dummy	Plate	Plate	Plate	Plate	Plate	Plate

Legend: Full Scale; Tunnel Measurements with Model Wake; Tunnel Measurements with Full Scale Wake (Post-trial)

Fig. 5.8 Scale effect due to propeller inflow condition

5.5.4   Direct Calculation of Bearing and Surface Forces

The overall calculation procedure for propeller loading calculation is as follows:

• Open water propeller flow analysis,
• Nominal wake harmonic analysis,
• Effective wake calculation—propeller/nominal wake interaction,
• Propeller cavity pattern simulation,
• Bearing forces/moments calculation, and
• Propeller induced hull pressure calculation.

In order to perform the aforementioned analyses, the following information is to be required:

• Measured nominal wake,
• Stern wave height,
• Water temperature,

- Ship speed,
- Propeller rpm,
- Draft,
- Propeller blade number,
- Propeller geometry (diameter, pitch, rake, skew angle, chord length, section profiles),
- Propeller location (related to ship hull), and
- Ship stern geometry.

5.6   Engine Excitation

The excitation forces and moments of low-speed diesel engine can be a significant source of hull vibration. Dominant forces acting on the engine structure originate from the combustion process (Guide Force Couples) and from the inertia of mass in motion (external forces). Guide force couples result from transverse reaction forces acting on the crosshead, causing rocking (H-couples), and twisting (X-couples) of the engine. These forces may cause resonance with the engine foundation structure. A possible solution for the eventual vibration problem may be installation of transverse stays (top bracings), connecting the engine's top structure to the ship hull.

Detailed consideration of the engine excitation forces and their effect on the ship hull structures is an important part in prevention of damaging structural vibrations. The magnitude of these forces and moments are to be obtained from the engine manufacturer.

5.7   Forced Vibration

5.7.1   General

Frequency response analysis is a method used to compute structural response to steady-state oscillatory excitation. The three-dimensional FE model is subject to systems of harmonic loads which represent the oscillatory excitation forces induced by propeller and main engine. Two different numerical methods can be used in frequency response analysis. A direct method solves the coupled equations of motion in terms of forcing frequency. A modal method utilises the mode shapes of the structure to reduce and uncouple the equations of motion. The solution for a particular forcing frequency is obtained through the summation of the individual modal responses. The choice of the method depends on the model size, number of the excitation frequencies and the frequency of excitation. If the modal method is used, all modes up to at least two to three times the highest forcing frequency are to be retained.

A frequency response is obtained by applying the cyclic propeller and engine forces and moments with varying frequencies, related to the shaft rpm and the number of blades per propeller and the order of main engine, to the model and solving the resulting dynamic problem given by the following equation of motion.

$$[M]\{\ddot{u}(t)\} + [C]\{\dot{u}(t)\} + [K]\{u(t)\} = \{F(t)\}$$

where

[C]   damping matrix
$\{\ddot{u}\}$ column matrix of accelerations
$\{\dot{u}\}$ column matrix of velocities
$\{u\}$ column matrix of displacements
$\{F\}$ column matrix of harmonic forces.

The structural response to this harmonic load is a steady state response at the frequency of the load. (It is assumed that the harmonic load continues long enough so that the transient response damps out). Therefore:

$$F(t) = Fe^{i\omega t}$$

$$u(t) = ue^{i\omega t}$$

Upon substitution, the equations of motion reduce to:

$$K + i\omega C - \omega^2 M]\{u\} = \{F\}$$

It is noted that both F and u are complex quantities due to damping C and to the fact that the various components of propeller and engine induced vibratory forces are not in phase with one another.

5.7.2   Critical Areas

In order to demonstrate the frequency response in terms of peak velocities within a certain structure without losing generality, a fairly large quantity of grid points within the structural model may be selected and their responses may be obtained in the form consisting of magnitudes and phase angles rather than complex components. Here is a list of the typical critical areas that are to be taken into consideration in any vibration analysis.

- Transom at centreline,
- Transom at port and starboard side,
- Point right above the propeller,
- Deck house:

- – Port and starboard side at bridge,
- – Centreline at bridge,
- – Port and starboard side at top of deck house (wheelhouse),
- – Centreline at top of deck house (wheelhouse),
- – Centreline at upper deck (forward and aft),
- Centreline at stern,
- One point on each cargo hold, on the transverse bulkhead,
- Engine room areas,
- Machinery areas, and
- Skeg area.

Depending on the ship type there may be additional areas of investigation. Care is to be taken to model the masses of the equipment. They are to be located on well-supported nodes as much as possible. A computer program may be used to post process the results obtained from the forced vibration analysis.

5.7.3   Damping

The total damping associated with overall ship hull structure vibration is generally considered as a combination of the following components:

- Structural damping,
- Cargo damping,
- Water friction,
- Pressure wave generation, and
- Surface wave generation.

For the forced vibration analysis, it is assumed that the effects due to structural damping, cargo damping, water friction and pressure wave generation can be lumped together. The effect of surface wave generation needs only to be considered for very low frequencies of vibration. This effect is generally neglected. For simplification, a constant damping coefficient of 1.5% of the critical damping may be used for the entire range of propeller rpm and main engine orders. Otherwise, more detail frequency-dependent damping coefficients may be used, if applicable.

Measurements

6

6.1 Introduction

6.1.1 Scope

These Guidance Notes provide general guidelines on the measurement and evaluation procedures for the vibration of hull girder and superstructure, local structures, main propulsion machinery and shafting, other machinery and equipment. This chapter covers overall procedures for acquiring, processing, and presenting data collected during sea trials based on ISO 4867 and SNAME T and R No. 2–29.

6.1.2 Application

The procedures provided are applicable to both turbine and diesel-driven vessels of all lengths. Actual items for measurements are to be determined as appropriate based on ship specifications or specific needs requiring the measurements.

6.1.3 Terminology

Average amplitude value: The value obtained from a Fast Fourier Transform (FFT) signal analyser, set in the averaging mode, equal to two times the value read form the amplitude spectrum, which is a rms level. This is equivalent to the average sinusoidal value obtained during steady speed trial conditions, and for design purposes, it is comparable to predicted response.

© The Author(s), under exclusive license to Springer Nature Switzerland AG 2025   65
F. Karkori, *Ship Vibration 1*, Synthesis Lectures on Ocean Systems Engineering,
https://doi.org/10.1007/978-3-031-75072-4_6

Free route: That condition achieved when the ship is proceeding at a constant speed and course with minimum throttle or helm adjustment.

Hull girder: The primary hull structure such as the shell plating and continuous strength decks contributing to flexural rigidity of the hull and the static and dynamic behaviour of which can be described by a free-free non-uniform beam approximation.

Hull girder vibration: That component of vibration which exists at any particular transverse plane of the hull so that there is little or no relative motion between elements intersected by the plane.

Local vibration: The dynamic response of a structural element, deck, bulkhead or piece of equipment which is significantly greater than that of the hull girder at that location.

Maximum repetitive amplitude (MRA): This is the largest repeating value of the modulated signal obtained on sea trial. It can be observed in the envelope of peak values on a time compressed strip chart of the modulated signal. This is the acceptance value specified in the codes.

Conversion factor (CF): The ratio of the MRA to the average amplitude obtained during steady speed trial condition.

"Peak hold": Most FFT analysers include a feature called "Peak Hold" which generates a spectrum of values which are the maximum spectral values in each frequency component for the set of all individual spectra comprising the "Peak Hold" spectrum. This is different from the MRA. In operation it represents an average value of a number of cycles of the signal evaluated rather than the maximum value of the single largest amplitude observed. The number of cycles averaged is based on the "window" or viewing time in seconds and, of course, increases with frequency.

RMS value: The root mean square value is the square root of the sum of the squares of all points of a signal. For a sinusoid, it is 0.707 times the single amplitude.

Severity of vibration: The peak value of vibration (velocity, acceleration or displacement) during periods of steady-state vibration, representative behaviour.

6.2   Instrumentation

6.2.1   General Requirements

Measurements are preferably to be made with an electronic system which produces a permanent record. The transducers may generate signals proportional to acceleration, velocity or displacement. Recording can be made either on magnetic tape, paper oscillographs, or in digital format. Use of paper oscillographs during the tests means that the vibration traces can be inspected directly and is extremely helpful in evaluating existing vibration problems. When displacement rather than either velocity or acceleration is recorded, the desired low-frequency signals associated with significant vibratory motion are the major components of a recorded trace. Thus, they are readily evaluated since they overshadow possible higher frequency signals with low displacement amplitudes. Provisions are to be made for suitable attenuation control to enable the system to accommodate a wide range of amplitudes.

An event marker is to be provided on the propeller shaft. Its position with respect to top dead centre of cylinder number one and a propeller blade is to be noted.

6.2.2 Calibration

Calibration procedures are categorised as system calibration or electrical calibration. In general, system calibration refers to a procedure that is done before installing instrumentation on board of the ship or as the transducers are installed. It is to be a complete reckoning of the sensitivity of the transducers, signal conditioning and recording equipment.

Electrical calibration refers to a procedure that can be accomplished usually at the recording centre, is considered a "spot check," takes only a few minutes, and can be done periodically during the vibration trials. Calibration procedures are different for different types of gauges and are discussed in this chapter.

6.2.2.1 Accelerometers

All accelerometers can be calibrated over the frequency range of interest by mounting on a shaker table or calibration device that is oscillating at know amplitudes. Normally, this is the type of system calibration that is used. Strain gauge and piezoresistive accelerometers can be calibrated for zero, ± 1 g by laying them on their sides, their bases and upside down, respectively. This provides a DC calibration only and is useful only if the conditioning and recording equipment operates at a frequency of zero Hz.

Once the transducers are installed, the "electrical calibration" is usually accomplished by an internal (to the amplifier) signal of known value being applied to the conditioning and recording equipment. In the case of strain gauge and piezoresistive accelerometers, a shunt resistor can be applied across one arm of the bridge and the value of the resistor can be equated to a certain acceleration. The latter results in a DC step being recorded.

6.2.2.2 Velocity Gauges

System calibration for velocity gauges is to be done on a shaker table that oscillates at known amplitudes and frequencies. There are no DC types of calibration suitable for these gauges. Electrical calibration is done by means of an internal signal of known value being fed into the conditioning and recording equipment. Since the conditioning equipment usually does not operate for DC signals, the known signal is usually a sinusoid.

6.2.2.3 Proximity Probes

Proximity probes should not be sensitive to changes in frequency and therefore can be calibrated for DC steps only. They are to be calibrated at several distances from the target by means of "feeler" gauges. Plastic feeler gauges are available that do not affect the signal and can be left in place while recording. Preliminary calibration can be done before installing the probes on board ship, but the final calibration is to be done with the probes in place because each target has a slightly different effect on the gauge. The only practical type of "electrical" calibration would be the substitution of a signal of known value.

6.2.2.4 Strain Gauges

The type of calibration used with strain gauges will depend on how they are used. Again, the signal output should not be sensitive to frequency and DC calibrations are adequate. If possible, the stain-gauged object is to be subjected to known loads and the resulting strains calculated and related to the signal output. If the object has a complex shape or if known loads are difficult to apply, the only choice is to accept the manufacturers' listed Gauge Factors and use a shunt resistor for calibration.

In most applications on board ship, the signal leads from the strain gauges are fairly long. This reduces the sensitivity of the gauges, requiring the shunt to be applied at the gauge rather than at the amplifier in order to obtain accurate results.

6.3   Measurement Conditions

Vibration tests are to be conducted under the conditions that are agreed by the shipyard and owner. The following measurement conditions are provided as a reference.

6.3.1 Environment Condition

6.3.1.1 Sea State

The test is to be conducted in a sea state no greater than the following as far as practicable:

- Sea state 1 for small craft,
- Sea state 2 for small ships (<10,000 tonnes), and
- Sea state 3 for large ships (>10,000 tonnes).

The actual sea state during the vibration measurement is to be clearly stated in the report.

6.3.1.2 Water Depth

The test is to be conducted in a water depth of not less than five times the draft of the ship, with machinery running under normal conditions. If a vessel normally operates in shallow water, the depth during the time of the test is to be representative of normal operating depths.

6.3.2 Loading Condition

The ship is to be ballasted to a displacement and trim as close as possible to the normal operating conditions within the ordinary ballasting capacity of the vessel. The draft aft is to ensure that the propeller is fully immersed.

6.3.3 Course

Tests are to be conducted for the following courses:

6.3.3.1 Free-Route Runs
During the free-route runs, the rudder angle is to be restricted to two degrees port or starboard. Rudder action is to be kept to a minimum.

6.3.3.2 Manoeuvres
Tests may also be conducted while completing the following manoeuvres:

- Hard turn port,
- Hard turn starboard, and
- Crashback.

In general, these tests are normally required for naval combatants or small craft.

6.3.4 Speed and Engine Power

Free-route runs are to be conducted at constant speeds from half shaft rotational speed or less (idle speed for diesels) to the maximum speed, using at least 10 equally spaced increments. Additional runs are to be made at smaller increments near each critical speed (the calculations will assist in locating criticals) and each service speed. If there are multiple shafts or propellers, all are to be run as close as possible to the same speeds (within two percent). In certain instances, it may also be appropriate to run with a single shaft or propeller.

Tests for hard turns to port and starboard are to be conducted at maximum speed. The crashback test is to be performed starting from the maximum speed and decreasing the propeller rpm at 10% of the maximum rpm per minute or less.

6.4   Measurement Locations

6.4.1   Stern

Vertical, transverse, and longitudinal measurements of the hull girder are to be taken at the centreline and as close as possible to the stern by means of a transducer. These measurements are to be for reference purposes. A pair of transducers for vertical vibration are also to be placed at the deck edge (one port, one starboard). They are to be located on the same deck and longitudinal location as the aforementioned transducer. It is to be ensured that the vibration of the hull girder is measured, excluding local effects.

6.4.2   Superstructure

Vertical, transverse, and longitudinal vibrations are to be measured at the following locations:

- Wheelhouse (navigation bridge), centreline at front of bridge,
- Bridge wing aft end,
- Main deck, centreline at front of deck house, and
- A pair of transducers to measure torsional motions of an aft deck house, when torsional vibration is to be determined.

Measurements are to be made in the range of 90% to 100% normal shaft rotational frequency.

6.4.3   Main Engine and Thrust Bearing

6.4.3.1   Geared Drive Turbines
Vertical, transverse and longitudinal measurements are to be taken on top of the thrust bearing housing. Recordings are also to be taken on one additional point on the thrust block foundation in the longitudinal direction. Measurements at other locations may be executed as optional. Other measurements to achieve the same results are permissible.

6.4.3.2   Direct Drive Diesels
Recordings are to be taken at the following locations:

- Top and foundation of the thrust bearing; if applicable,
- Top forward end of the main engine, in the longitudinal direction,
- Top forward and aft ends of the main engine, in the vertical and transverse directions,

- Forward end of the crankshaft, in the longitudinal direction, and
- Forward and aft ends of the engine foundations, in the vertical and transverse directions.

For the other measuring points, optional recordings may be taken at constant shaft rotational frequency. Measurements are to be made throughout the normal operational range of the ship.

6.4.4 Lateral Shaft Vibration

If measurement is conducted, vertical and transverse vibration measurements are to be made on the shaft relative to the stern tube. Additional measurement points are optional and may be taken. These measurements are to be made throughout the normal operational range of the ship. In order to eliminate error, shaft run-out is to be checked by rotating the shaft by the turning gear and recording the first-order signal. This signal is to be phased and the shaft vibration measurement corrected accordingly.

6.4.5 Torsional Shaft Vibration

Torsional vibration measurements are to be taken either at the free end of the propulsion machinery, using a suitable torsional vibration transducer, and/or on the main shafting, using strain gauges. Alternatively, depending on the system characteristics, a mechanical torsiograph, driven from a suitable position along the shafting or free end, may be used for this purpose.

6.4.6 Local Structures

If severe local vibration exists during sea trials, vertical, transverse, and longitudinal measurements are to be taken at the suspect location. This is necessary in order to determine the need for corrective measures.

6.4.7 Local Deck Transverse

Vertical and transverse bending vibration measurements are to be taken at the deck edge using a significant number of points necessary to determine the mode shapes at low frequencies while avoiding local resonances. These types of measurements are to be made by use of a reference transducer at the stern along with a portable transducer. Torsional modes may require phased deck-edge measurements.

6.4.8   Local Machinery and Equipment

Vertical, transverse and longitudinal measurements are to be taken at the outside of machinery where there is evidence of large vibration amplitudes. Types of machinery and equipment include diesel generator, electric motors, air compressors, ballast pumps, etc. Vibration measurement is to be taken at the bearing position.

6.4.9   Shell Near Propeller

If necessary, the measurements of hull surface pressure are to be taken in order to confirm design estimates, to obtain design data or to investigate potential cavitation problems. The pressure transducers are to be located as follows:

- Two transducers in the propeller plane, and
- One transducer at 0.1 D forward of the propeller plane.

To minimise the effect of plate vibration, all transducers in the hull plating are to be located as close as possible to adjacent frames or partial bulkheads.

6.5   Data Processing Analysis

6.5.1   Measured Data

6.5.1.1   Displacement, Velocity, and Acceleration

The type of data that is to be recorded (displacement, velocity or acceleration) may depend on the appropriate frequency range. For example:

For frequencies	Measure
Below 300 rpm	Displacement
Between 300 rpm and 6,000 rpm	Velocity
Above 6,000 rpm	Acceleration

If a particular data type is critical in determining damage to machinery or structure or for human response, the appropriate data type may be used in-lieu of that above. For example:

For	Measure
Mechanical or structural fatigue	Displacement
Human response	Velocity
High frequency machinery components	Acceleration

6.5.1.2 Frequency Range

The frequency ranges of transducers, signal conditioners and recording equipment are to be chosen to match the frequency components of interest in the data being recorded. Frequencies that are known to be present but not of interest are to be excluded.

Do not record unwanted frequencies. This will make it difficult to separate the level of the frequency components of interest from the "noise." Also, do not choose a frequency analyser with a broader range than necessary. This will decrease the resolution and accuracy of the results (Table 6.1).

6.5.1.3 Time History or Frequency Spectra

In case where the vibration is self-excited, such as rotating machinery, obtaining the frequency spectra directly is appropriate. In this case, amplitudes are fairly constant, and a large sample is not required. However, most vibration data gathered from ships are excited by the sea, propeller or some combination (e.g., hull girder vibration). For this type of data, the amplitudes are modulated and have a randomness associated with them and may require a sample large enough to account for the randomness, which for blade frequencies and its harmonics are to be three minutes.

6.5.2 Performance of Measurements

- For the free-route run, data are to be recorded for at least three minutes after steadying at each ship speed, or for at steadily increasing shaft speed less than 1 rpm per five minutes.
- For the hard turns to port and starboard, recording is to commence before the turn and continue until maximum vibration has passed (when settled into the turn).

Table 6.1 Typical frequencies ranges

Item	Frequency range
Ship motion	Below 1 Hz
Major structure (deckhouses, masts, etc.)	Between 1 to 10 Hz for the lower modes
Propeller shaft rotation, gear frequencies, and blade rate	To be determined from the machinery characteristics

- For the crashback test, data are to be recorded before the propeller rpm is reduced and continue until the ship comes to rest.

6.5.3 Analysis Methods

The methods of analysing shipboard vibration data are presented in SNAME T and R No. 2–29. The manual method is to be used by an engineer during sea trial in order to complete a quick assessment of test results and to identify potential problems. For the final test report, the use of the fast digital electronic (FFT) analysers is to be employed. For more details, refer to SNAME T and R No. 2–29.

6.5.3.1 Manual

The manual method of analysis involves measuring frequencies and amplitudes of vibration components from an oscillograph record. The analysis of periodic waveforms is based on the principle that any periodic record is a superposition of sinusoids having frequencies that are integral multiples of the lowest frequency present. The lowest frequency is determined by the smallest portion of the record that repeats itself, or one cycle.

Ship vibration often involves more than one component, so that decomposing the signal into its separate components would mean discerning beating and modulating and applying the appropriate analysis techniques. Thus, having magnetic tape of the actual recorded signal available for detailed analysis, as applicable, is highly recommended.

6.5.3.2 Envelope

This method of analysis consists of filtering the signals to obtain the MRA of a narrow frequency band of interest and recording the result on a chart recorder at slow play out speed. This condenses a several minute record into an envelope just several inches long, from which the MRA can be visually noted. Care must be taken that the filter used does not pass a significant amount of any component other than that of the component of interest. For this reason, it is important to use precision filters.

This method is the most direct approach to the determination of the MRA of individual components. If the MRA of the multi-frequency amplitude is required, the desired response would be obtainable by simply recording the real-time record at a slow speed and measuring the overall amplitude and multiplying by the sensitivity of the measuring/recording system. However, for obtaining the amplitudes of individual components for use in evaluating design analyses and/or developing improved crest values to be used in conjunction with spectral analyses, it will be necessary to include filtering systems.

6.5.3.3 Spectral

Two common types of spectra that can be found with most analysers are the "average" and the "peak" amplitude. For both types, the record is broken down into segments for frequency analysis. The segments vary in length with the frequency range of the analysis, but for ship vibration studies they would be several seconds long. For each segment, the analyser finds the rms level of each frequency component (i.e., in each frequency interval, or for each "line"). The number of lines (resolution) varies with the analyser, but most often falls between 100 and 1,000. Ship vibration records will normally be one or several minutes long and will contain many segments. If the "average" spectrum is chosen, the analyser will average all the rms levels found in like frequency intervals and produce the "average" spectra as "average rms" values. This value will correspond to the rms value shown in the ISO-6954 formula:

$$(MRA) = \left(C_F \sqrt{2}\right) \times \text{ rms value}$$

where C_F denotes "conversion factor" and $C_F \sqrt{2}$ denotes "crest factor." The average value, or average amplitude is equal to $\sqrt{2}$ times the "average rms" value read from the "average" spectrum.

This is equal to the average sinusoidal value obtained during steady speed trial conditions and, for design purposes, is comparable to predicted response.

The term "peak" is frequently misinterpreted when used in conjunction with the application of the spectral analyser in the evaluation of shipboard hull vibration. To clarify the use of the term, note that the peak value of any vibration cycle is the difference between the mean and maximum values of that cycle. However, in the case of a modulating signal we note that the "peak" value increases to a "maximum" value, every few cycles. Thus, the "peak" of the modulating signal represents the maximum repetitive value or MRA and is actually equivalent to the peak value of each segment and corresponds to "peak hold" for that segment.

When choosing the "peak" spectrum, the largest rms values are obtained in each frequency interval found among several segments. These values are called "peak rms" values with some variation in results, depending on the frequency range used. A higher frequency range will involve broader frequency intervals and yield higher results. The average spectral value will always be less than the MRA. The relationship between the latter two will depend on the rate of modulation compared to the length of the segment. If the modulation is slow, the amplitude will be near its maximum for the entire length of some segments and the two will be close. If the modulation is fast, the peak will be closer to the average amplitude.

The amplitude variation at blade-frequency between "average spectral," "peak spectral" and MRA (derived by the envelope method) is shown in Table 6.2 for example. The data represents the average of three full-power runs measured at 114 rpm, and shows alternating pressure measured over the propeller in psi, and alternating displacement in mills at both the stern of the hull in the vertical direction and at the hub of the bull gear in the longitudinal direction. This data indicates that the conversion factor (C_F) of 1.8 and the crest factor of 2.5 are satisfactory for these trials, particularly in regard to the

Table 6.2 Examples of alternate vibration measurements

	Avg spect (rms)	Peak spect (rms)	MRA (Envelop)	Crest fctr CF 2	Conv fctr CF
Hull pressure, psi	1.3	1.65	2.65	2.04	1.45
Stem vertical disp., mils	1.7	0.329	5.80	3.41	2.42
Bull gear hub, long disp., mils	1.87	3.22	4.79	2.56	1.81
Average					1.89

displacement amplitudes obtained on the hull and bull gear measurements. It also indicates that the hull and machinery vibration measurements, on which the acceptance criteria are based, supports the position that the peak spectral measurements will be significantly lower than the "true" MRA obtained by the envelope method of analysis.

6.5.3.4 Histograms of Instantaneous Values

Some analysers sample the signal and obtain histograms of the instantaneous values. This capability might be useful with ship vibration records, but there are several considerations. First, we are usually concerned with the amplitudes of blade-frequency and its harmonics, so that filtering is necessary as it is in the envelope method. The limitations and necessary steps in the use of the filtering of the higher harmonics will also apply to the statistical method.

Second, we are concerned with the MRA of the major frequency components. The histograms are usually obtained by sampling all the points on the record, not just the peaks. The amplitude, which is exceeded by only one or two percent of the samples, would involve only the tips of the largest cycles and may be comparable to the MRA. The amplitude, which is exceeded by some percentage of the samples, could be determined from the cumulative probability plot.

6.5.3.5 Histograms of Peak Values

This procedure would be closer to the envelope method and can be obtained with an analogue to digital (A/D) converter and a microcomputer in conjunction with filters. It entails sampling the filtered signal, as described for the histograms of instantaneous values, then finding the peak (and trough) of each cycle. The peak amplitudes are then put into a histogram or cumulative probability plot. The top three, five, or ten percent of the peak amplitudes may be comparable to the MRA. Although it is not likely that this procedure is currently being used for routine analysis of ship vibration data, it would seem to be the most accurate of those discussed. However, considering the vagaries in the total scope of effort in the measurement and analysis of ship vibration, this degree of complexity would not be warranted.

6.6   Measurement Report

6.6.1   Analysis and Reporting of Data

6.6.1.1   Analysis

The contents of analysis are to be selected from the following depending on the necessity:

- Severity of vibration at the propeller shaft rotational frequency for hull girder transducers,
- Severity of vibration at blade rate frequencies for hull girder and machinery transducers,
- Severity of vibration of each detectable harmonic of shaft rotational frequency or blade rate for hull girder and machinery transducers, as applicable. Also, the severity of each detectable multiple of crankshaft rotational frequency in the case of geared diesel installations,
- Severity of local structural vibration at all measurement locations,
- Mode shape of local vibrations. Use hull girder vibration as reference for the mode shape,
- Severity of vibrations of local machinery or equipment at all measurement locations,
- Phase relation between various transducers at blade rate, as applicable, using a suitable reference datum (e.g., a hull girder or machinery transducer or event marker),
- For diesel engines, phase relation is to be provided between all transducers measuring in the longitudinal direction and for the transducers on top of the engine measuring torsional motions; therefore, each group is to be measured simultaneously, and
- Severity of vibration at hull girder and machinery resonances.

Note: If beating exists, it is to be noted by recording the maximum and minimum values of the amplitude and the frequency of the beat.

6.6.1.2   Reporting of Data

The contents of reporting are to be selected from the following depending on the objectives of the report:

- Particulars of test ship,
- Particulars of propulsion-shaft system,
- Particulars of main diesel engine or turbine driven plants,
- Provide sketch of inboard profile of hull and superstructure,

- A lines plan,
- A midship section,
- A sketch of screw aperture,
- A sketch showing locations of hull girder and machinery transducers and their directions of measurement. Transducer locations for local vibration measurements are to be shown on a separate sketch where the precise location of the transducer is noted,
- Trial conditions during vibration measurements including sea state, wave direction, water depth, draft at FP and AP, propeller immersion,
- Plots of displacement, velocity or acceleration amplitudes versus speed for shaft rotation frequency, blade rate, or machinery excitation frequency or any harmonic thereof,
- Profiles of local deck vibration at each resonance from port to starboard and from the nearest aft to the nearest forward structural bulkhead,
- Tables of all significant vibration severities and their location and frequency. Include the shaft rotational frequency for machinery-excited vibration,
- Hull girder natural frequencies identified from stern measurements and any unusual vibration condition encountered,
- Results of vibration measurements at local areas,
- Results of vibration measurement during manoeuvres,
- Longitudinal vibration of the propulsion system during manoeuvres,
- Method of analysis of results, and
- Type of instrument used.

The report is to note the hull natural frequencies and modes which have been identified. Also, the report is to address any undesirable or unusual vibration condition encountered.

7.1 Introduction

Criteria for assessing the acceptability of the vibration levels from vibration measurements on sea trials have matured substantially in recent years. By incorporating the international standards and industry practices, this chapter provides the vibration acceptance criteria covering three areas as a reference:

(1) Vibration limits for crew and passengers,
(2) Vibration limits for local structures, and
(3) Vibration limits for machinery.

Vibration acceptance criteria are usually defined in ship specifications. The actual vibration acceptance criteria are to be determined as appropriate based on the ship specifications and manufacturer's specifications.

7.2 Vibration Limits for Crew and Passengers

7.2.1 Criteria for Crew Habitability and Passenger Comfort

The classification society guides for crew habitability and passenger comfort provide acceptance criteria based on the evaluation of both sea-motion induced and mechanically-induced vibrations for the purpose of granting ABS optional notations. The acceptance criteria are stated in terms of overall frequency-weighted rms acceleration values from 0.5 to 80 Hz. Motion sickness criterion is stated as a Motion Sickness Dose Value $MSDV_Z$ for frequencies between 0.1 and 0.5 Hz.

F. Karkori, *Ship Vibration 1*, Synthesis Lectures on Ocean Systems Engineering,
https://doi.org/10.1007/978-3-031-75072-4_7

Table 7.1 Maximum weighted RMS acceleration levels for crew habitability

Class optional notation	Frequency range	Acceleration measurement	Maximum level
HAB	0.5–80 Hz	a_w	0.47 m/s^2
HAB +	0.5–80 Hz	a_w	0.315 m/s^2

Table 7.2 Maximum weighted RMS acceleration levels for passenger comfort

Class optional notation	Frequency range	Acceleration measurement	Maximum level
COMF	0.5–80 Hz	a_w	0.315 m/s^2
COMF +	0.1–0.5 Hz	$MSDV_Z$	30 m/s$^{1.5}$
	0.5–80 Hz	a_w	0.20 m/s^2

The criteria are based on BS 6841 (1987) and ISO 2631 (1997). The frequency weighting curves in BS 6841 (1987) are used for the vertical axis (W_b filter for x-axis). The horizontal axes use the frequency weighting curves in BS 6841 and ISO 2631–1 (W_d filter for y- and z-axes). For motion sickness dose value computation, a different frequency weighting curve from BS 6841 is applied (W_f for z-axis).

The maximum weighted rms acceleration level for crew habitability is shown in Table 7.1. The class criteria require each single axis as well as the combined multi-axes levels be less than or equal to the values expressed in Table 7.1. Note that the low frequency range below 1 Hz is included to account for the human response to sea conditions including heave, slamming and whipping. The maximum weighted rms acceleration level for passenger comfort is shown in Table 7.2. Again, both single axis and multi-axes levels must be less than or equal to the values expressed in Table 7.2. Note that the MSDV (Motion Sickness Dose Value) based on vertical motion for the frequency range of 0.1 to 0.5 Hz is also included in Table 7.2.

Most classification societies provide optional classification notations for crew habitability (HAB/HAB+) and passenger comfort (COMF/COMF+). For details on the acceptance criteria and measurement procedure, refer to the appropriate class guide for passenger comfort on ships and class guide for crew habitability on ships. Where applicable, readers may also refer to the class guide for crew habitability on offshore installation for offshore applications.

7.2.2   ISO 6954 (1984) Criteria for Crew and Passenger Relating to Mechanical Vibration

ISO 6954 (1984) has also been widely used as acceptance criteria for crew habitability and passenger comfort. The criteria are designed to ensure vibration levels are below at which crew and passenger do not experience discomfort. ISO 6954 criteria are shown in Fig. 7.1, which can be transformed into the following:

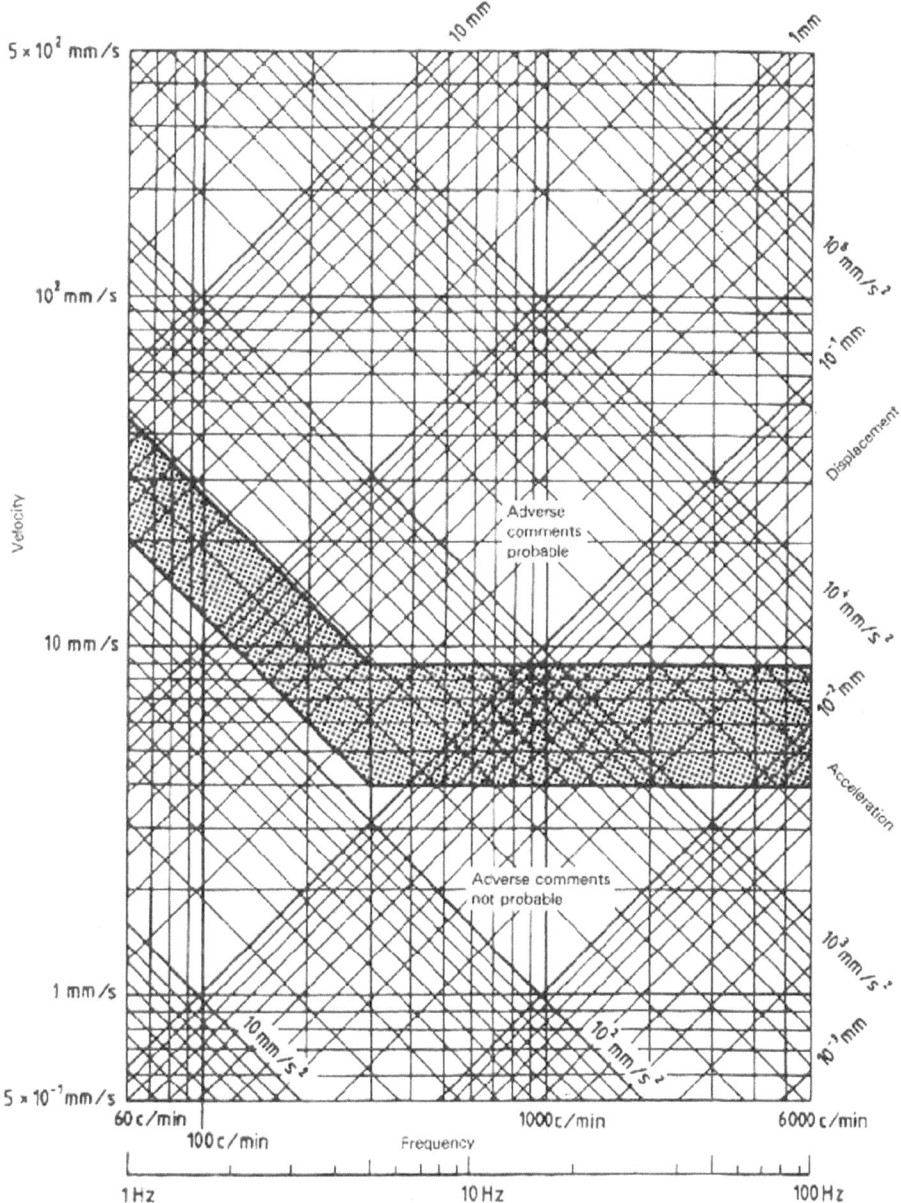

Fig. 7.1   ISO 6954 (1984)

- For each peak response component (in either vertical, transverse, or longitudinal direction), from 1 to 5 Hz, the acceleration is acceptable below 126 mm/sec^2, and adverse comment is probable above 285 mm/sec^2,
- For each peak response component (in either vertical, transverse, or longitudinal direction), from 5 Hz and above, the velocity is acceptable below 4 mm/sec, and adverse comment is probable above 9 mm/sec.

Criteria in the format of Fig. 7.1 characterises ship vibration as a simple harmonic (i.e., periodic at a single frequency). The criteria are readily applied to the evaluation of FE-based vibration analysis, as described in Chap. 5. However, the ship vibration in seaways is actually random in nature (i.e., it is composed of components at all frequencies rather than at a single one). The random character of ship vibration is clearly evident in records from underway vibration surveys. Note that ISO 6954 (1984) criteria are given for the peak values considering the modulation of ship vibrations in seaways. For the measurement and evaluation of ship vibration in seaways, the Envelope method in Chap. 6 would be the most direct and simple method to effectively measure the peak values.

7.2.3 ISO 6954 (2000) Criteria for Crew and Passenger Relating to Mechanical Vibration

ISO 6954 (1984) has been revised to reflect recent knowledge on human sensitivity to whole-body vibration. The frequency weighting curves are introduced to represent human sensitivity to multi-frequency vibration for a broad range of frequencies, which are consistent with the combined frequency weighting in ISO 2631–2. ISO 6954 (2000) provides criteria for crew habitability and passenger comfort in terms of overall frequency-weighted rms values from 1 to 80 Hz for three different areas. The simplified presentation is shown as Table 7.3.

Table 7.3 Overall frequency-weighted RMS values (ISO 6954: 2000)

Class optional notation	Area classification					
	A		B		C	
	mm/s^2	mm/s	mm/s^2	mm/s	mm/s^2	mm/s
Values above which adverse comments are probable	143	4	*214*	6	286	8
Values below which adverse comments are not probable	71.5	2	*107*	3	143	4

Note: The zone between upper and lower values reflects the shipboard vibration environment commonly experienced and accepted

Area classification

A: Passenger Accommodations,
B: Crew Accommodations,
C: Workspaces.

7.3   Vibration Limits for Local Structures

Excessive ship vibration is to be avoided in order to reduce the risk of structural dam-age on the local structures. Structural damage such as fatigue cracking due to excessive vibration may occur on local structures, including but not limited to engine foundation structures, engine stays, steering gear rooms, tank structures, funnels, and radar masts. It should be noted that the structural damage due to the excessive vibration vary according to the local structural detail, actual stress level and local stress concentration and material property of the local structures. Therefore, the vibration limits for local structures are to be used as a reference to reduce the risk of structural damage due to excessive vibration during the normal operating conditions.

Figure 7.2 is the vibration limits for local structures, often adopted in marine industry, below which the risk of fatigue cracking due to vibration is generally expected to be low. The heavy lines are the vibration limits for local structures; recommended below the lower limit, and damage probable above the upper limit, with a "grey area" in between. Above 5 Hz, the vibration limits are specified in terms of velocity amplitude, and below 5 Hz in terms of displacement. The local structure vibrations of main interest are generally above 5 Hz. The vibration limits can be transformed into the following:

- For each peak response component (in either vertical, transverse, or longitudinal direc-tion), from 1 to 5 Hz, the displacement is recommended below 1.0 mm, and damage is probable above 2.0 mm, and
- For each peak response component (in either vertical, transverse, or longitudinal direc-tion), from 5 Hz and above, the velocity is recommended below 30 mm/sec, and damage is probable above 60 mm/sec.

It is noted that the simple vibration limits described above may not be applicable to all the local structures with different structural configurations and details. In the case of tall structures and/or softly shaped structures such as masts, for example, the actual stress level due to vibration is generally small and the vibration limits of 1.0 or 2.0 mm may be too conservative. On the other hand, in the case of local stiffened panels with fixed ends, for example, the vibration limits of 1.0 or 2.0 mm may be less conservative. Therefore, the application of vibration limits for specific local structures may vary depending on the vessel specification agreed by shipyards and ship owners.

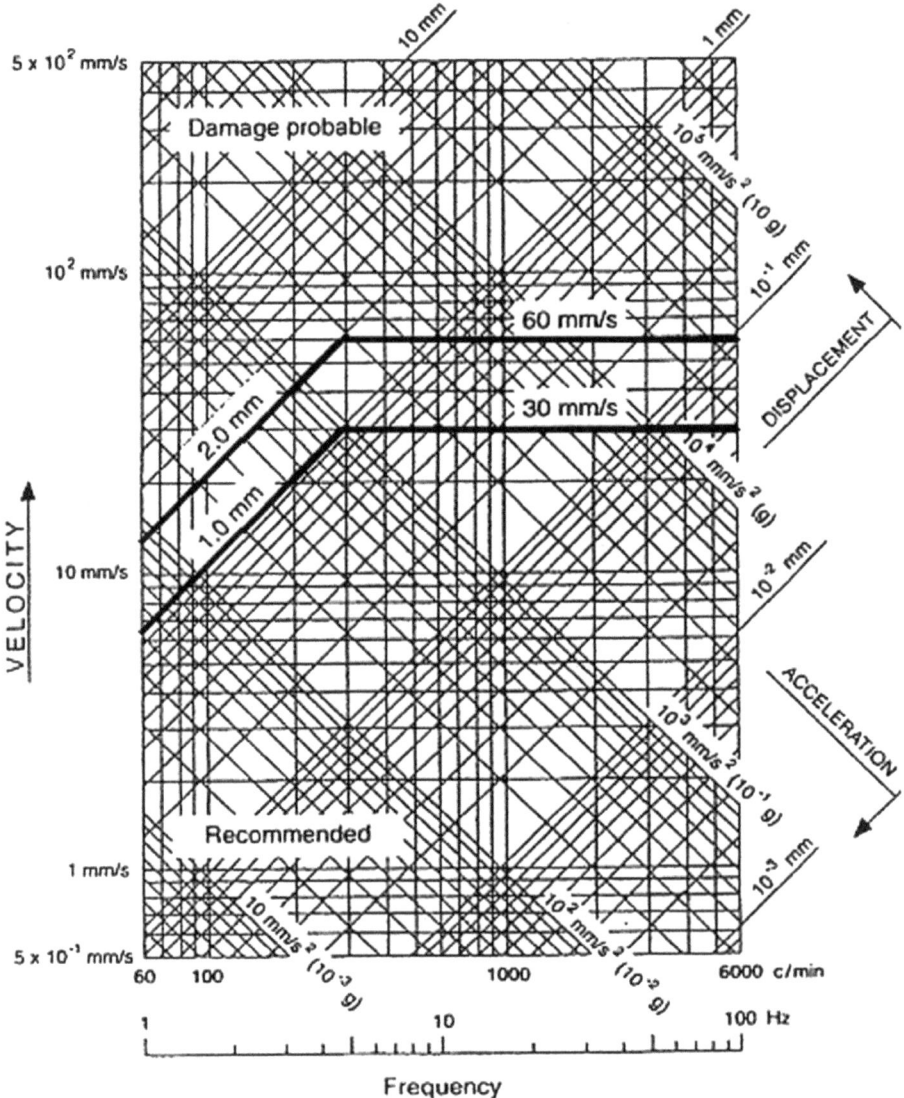

Fig. 7.2 Vibration limits for local structures

7.4   Vibration Limits for Machinery

7.4.1   Main Propulsion Machinery

The vibration of main propulsion machinery tends to be severe because of the excitation from the propellers. The machinery of primary concern is to be the main propulsion machinery, principally in longitudinal vibration at propeller blade rate frequency (see Chap. 4). The vibration criteria for the main propulsion machinery are to be provided from manufacturers. Otherwise, when the data on the vibration criteria are not available, the following criteria are recommended as a reference.

ANSI S2.27 (2002) and SNAME T & R 2-29A (2004) provide comprehensive guidelines on the vibration limits for the main propulsion machinery. The vibration limits are provided in terms of broadband rms values with multi-frequency components (nominally from 1 to 1,000 Hz). The longitudinal vibration (rms, free route) at thrust bearing (and bull gear hub for geared turbine drives) is to be less than 5 mm/s rms. For other propulsion machinery components exclusive of engines, propellers and shafting aft of the thrust bearing, the longitudinal vibration is to be less than 13 mm/s rms. For stern tube and line shaft bearing, the lateral vibration is to be less than 7 mm/sec rms. For direct diesel engines (over 1,000 HP, slow and medium speed diesels connected to the shafting), the vibration limits are 13 mm/sec at the bearings and 18 mm/sec on the engine tops, in all three directions. For high-speed diesel engines (less than 1,000 HP), the vibration is to be less than 13 mm/sec at the bearings and engine tops, in all directions (Table 7.4).

A primary concern in the longitudinal propulsion machinery vibration is the avoidance of thrust reversal at the thrust bearing due to system longitudinal resonance, which can result in destruction of the thrust bearing. ANSI S2.27 provides the criteria on the alternating thrust at thrust bearing. During free-route runs, the broadband peak value of the alternating thrust on the main thrust bearing is to be less than 75% of the mean thrust for that speed, or less than 25% of the full power mean thrust, whichever is less.

As shown in Chap. 4, Sect. 7.2, it may not be possible to avoid a longitudinal resonance in the operating range with long shaft ships. A critical at least 20% below the full power

Table 7.4 Vibration limits for main propulsion machinery

Propulsion machinery	Limits (rms)
Thrust bearing and bull gear hub	5 mm/s
Other propulsion machinery components	13 mm/s
Stern tube and line shaft bearing	7 mm/s
Diesel engine at bearing	13 mm/s
Slow and medium speed diesel engine on engine top (over 1000 HP)	18 mm/s
High speed diesel engine on engine top (less 1000 HP)	13 mm/s

rpm is the usual practice. Even if the critical occurs at an rpm less than 20% of full power, it may be very difficult to achieve the criteria proposed by ANSI S2.27 at the resonant point, depending on the blade-rate alternating thrust amplitude and the damping present. In this case, an rpm band around the critical might have to be restricted for operation as the only reasonable recourse.

7.4.2 Machinery and Equipment

The vibration criteria of machinery and equipment are to be provided by manufacturers. Otherwise, when the data on the vibration criteria are not available, the following criteria are recommended as a reference in terms of overall rms value (nominally from 1 to 1,000 Hz) for normal operating conditions.

- For reciprocating machinery, the vibration in all directions is to be less than 10 mm/ sec rms at bearings, and
- For rotating machinery, the vibration in all directions is to be less than 9 mm/sec rms at bearings.

The machinery includes but is not limited to generators, motors, centrifugal pumps, compressors, turbochargers, blowers and fans. The application of vibration limits may vary depending on specific type, size, configuration, and mounting of the machinery.

ISO 10816 provides guidelines on the vibration criteria in terms of overall rms values (from 2 to 1,000 Hz) for the non-rotating and, where applicable, non-reciprocating parts of general machines, measured at the bearings or bearing housings. It is noted that the criteria relate only to the vibration produced by the machine itself and not to vibration transmitted to it from outside. ISO 10816 is complemented by ISO 7919, which provides guidelines on the vibration criteria for the rotating parts of the machines.

In accordance with the applicable class rules for the vessel, a class recognised condition monitoring company may be required to submit the measured machinery vibration data and acceptance criteria to class. The vibration limits for machinery given in this chapter may be used as a reference for the acceptance criteria of class approved condition monitoring programmes, if applicable.

8.1   Corrective Investigations

The approach to resolving ship vibration problems, as with most engineering problems, involves two steps. The first step is to clearly establish the cause of the problem, and the second step is to implement the changes required to eliminate it in an efficient manner.

In about 80 percent of cases the basic cause of a ship vibration problem is its propeller. Whether or not the vibration of a particular ship has its source in the propeller is easily established from underway vibration measurements. If at some shaft rpm the measured frequency of the vibration is predominantly rpm times propeller blade number, and varies directly with shaft rpm, then the propeller is definitely the exciting source. If blade-rate frequency, or its multiples, is not strongly detectable in the records, then it is almost certain that the propeller is not the primary excitation, unless the records exhibit a strong shaft-rate frequency, which could indicate propeller mechanical or hydrodynamic unbalance difficulties, but these are rather rare.

Once the excitation frequency has been established from the underway measurements, next in order is to establish whether or not resonance with structural natural frequencies plays a significant role in the magnitude of the vibration. For non-cavitating propellers, excessive hull vibration is expected to be resonant vibration. Resonant vibration is established by varying shaft rpm in steps and recording vibration amplitude successively at each rpm in the region where the problem has been identified as being most intense. If a plot of displacement amplitude versus rpm shows a definite peak with increasing rpm followed by decline, then resonant vibration is established, and the position of the peak establishes the natural frequency of the resonant structural mode. If the amplitude/rpm characteristic does not peak but has an increasing trend as roughly rpm squared in the upper power range, then structural resonance is not playing a major role.

© The Author(s), under exclusive license to Springer Nature Switzerland AG 2025   87
F. Karkori, *Ship Vibration 1*, Synthesis Lectures on Ocean Systems Engineering,
https://doi.org/10.1007/978-3-031-75072-4_8

If, alternatively, the amplitude/rpm characteristic increases very rapidly only in the immediate vicinity of full power without establishing a definite peak up to the maximum obtainable rpm, a full power resonance may or may not be indicated. This exhibition can be entirely the manifestation of the onset of propeller intermittent sheet cavitation, which tends to produce an almost discontinuous amplification of the hull surface excitation at the onset rpm. The sudden appearance of strong higher harmonics of blade-rate frequency in the vibration records, accompanied by violent pounding in spaces above the counter, are good indications of a full-power non-resonant vibration problem caused by excessive propeller cavitation.

If non-resonant vibration due to propeller cavitation is established, then the underway survey could probably be discontinued, with attention then turned to hydrodynamic design changes in the stern/propeller configuration. This course of action is considered in the next subsection.

If the problem is established as highly localised resonant vibration of plating panels, piping, and the like, then the vibration survey likewise need go no further. It is usually quite obvious in such cases how natural frequency changes, through local stiffening, can be effectively and expediently accomplished to eliminate the locally resonant conditions.

If, on the other hand, the vibration problem is established as a resonant condition of a major substructure, such as a deckhouse, which has been all too often the case, then the vibration survey had best proceed to obtain mode shape information in the interest of an expeditious correction programme.

8.2   General Approach

Just as in developing a vibration-sufficient ship design, there exist three possibilities for correcting a vibration-deficient one in normal practice:

(1) Reduce vibratory excitation,
(2) Change natural frequencies to avoid resonance, or
(3) Change exciting frequencies to avoid resonance.

Most of the excessive diesel engine excited hull vibration can usually be corrected by the following provisions:

(1) Mechanical or electrical moment compensators against engine external moments, or
(2) Engine lateral stays of weld, hydraulic or friction type against engine lateral vibration, or
(3) Axial damper against crank shaft vibration, or
(4) Dynamic absorber against fore-and-aft vibration of superstructure.

The detail information necessary for the installations of above provisions is to be provided by engine or equipment manufacturers. Otherwise, the achievement of any of the three correction possibilities identified above may involve modifications in either stern/propeller hydrodynamics or hull structure.

8.3 Hydrodynamic Modifications

The most effective way to reduce propeller vibratory excitation is to reduce the circumferential nonuniformity of the hull wake in which the propeller operates, as discussed in detail in Chap. 3. In the early design stage, acceptable wakes can be achieved by taking proper care with stern lines. (Refer to Chap. 3, Sect. 8.3 for the specific guidelines). In a post-design corrective situation, basic lines changes are, of course, not possible. However, with good luck in the case of poor stern lines, considerable improvements in wake might be accomplished by back fitting one of the several types of wakes, adapting stern appendages. The partial tunnel Fig. 8.1, has been the most universally applied of the wake adapting appendages, which also include vortex generators and wake adapting propeller ducts.

The idea is to divert the upward-aft flow along the buttock lines forward longitudinally into the upper propeller disc to reduce the wake spike near top-dead-centre. This device will work most effectively on the buttock-flow type of stern; the partial tunnel was applied successfully over the years on Great Lakes ore carriers, most of which have

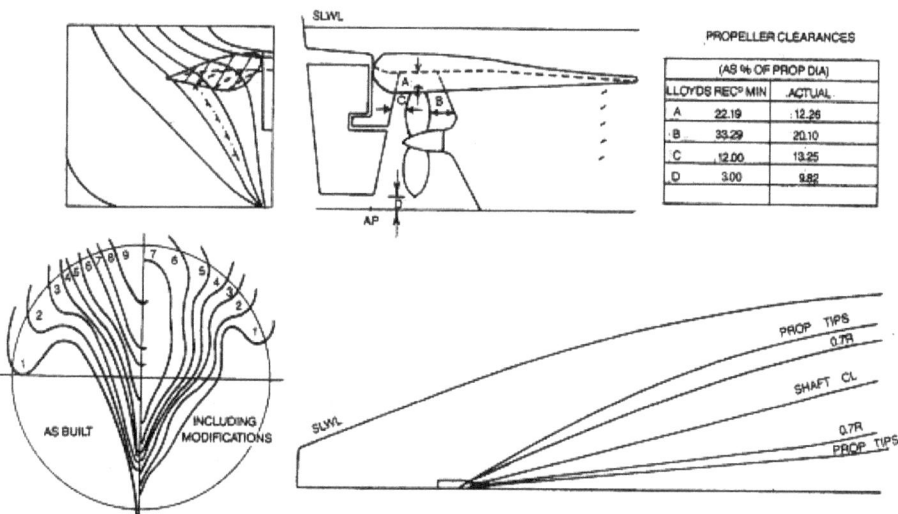

Fig. 8.1 Wake improvement with special lines-adapting stern devices conventional stern cargo ship

barge-type sterns with very steep buttock angles. On the other hand, for sterns which exhibit a waterline-flow character, the partial tunnel would be expected to be more or less ineffective due to the lack of upward flow to divert. However, the effectiveness of the partial tunnel cannot always be accurately judged by simply classifying a prospective application as one of the two limiting cases of buttock versus waterline flow. For example, the stern shown on Fig. 8.1, from Rutherford (1978–79) might be classified as more of a waterline flow, yet the modifications shown produced significant improvement in the nominal wake, as exhibited by the before and after axial velocity contours.

The contour lines on Fig. 8.1 are the lines of constant wake fraction, so that 9 is 0.9, or 10% of the axial free stream velocity, which represents a bad wake. The Fig. 8.1 modifications do however include vortex generators as well as the partial tunnel, and the contributions of each to the wake improvements shown are not known.

The decision to retrofit a wake adapting stern appendage should not be made lightly without quantification of the advantages and disadvantages. A price is paid for appendages in increased hull resistance. As a minimum, model tuft-tests with and without the appendage is to be performed to observe the change in stern surface flow. The absence of any noticeable smoothing may be misleading; however, a wake survey can show improvements in the propeller plane not discernible in the tuft behaviour. Furthermore, aside from nominal wake considerations, it has been found that greatest wake improvements are sometimes achieved through propeller/appendage interaction. This implies that model tuft-tests are to be conducted both with and without the operating propeller. In these cases, the best indicator of significant effective wake improvements from the standpoint of vibratory excitation may be an improvement by several percentage points in the propulsive efficiency from the model SHP test conducted with and without the wake adapting appendage. This is as explained in Hylarides (1978).

Aside from wake improvements, the only recourse for reducing propeller excitation is modification or replacement of the propeller. Some instances of successful modifications of troublesome propellers have been reported. For example, trimming blade tips by several centimetres to reduce wake severity at the extreme propeller radii can produce improvements, but some degree of rpm increase must then be tolerated. Successful modifications of existing propellers are rare because of the usually unacceptable trade-offs of performance degradation against vibration improvement. The same disadvantages exist in propeller replacement considerations. Replacement propellers, with modified features, such as changed blade number, reduced diameter (for increased hull clearance), increased blade area, reduced pitch in the blade tips, etc., may relieve the vibration problem, but often for a dear price in vessel performance. It is unfortunate that, with the exception of moderate blade skew, all of the measures available in propeller design for reducing vibratory excitation, once the stern lines are established, act also to reduce propeller efficiency. It cannot be emphasised strongly enough that the greatest insurance against propeller induced vibration problems is to place high emphasis on wake uniformity in making trade-offs in the original establishment of vessel lines at the concept design level.

8.4   Structural Modifications

The most cost-effective approach for eliminating structural resonances is usually to shift natural frequencies through structural modifications. The alternative is to shift exciting frequency by changes in engine rpm or number of propeller blades.

Just as with hydrodynamics related problems, the most intelligent way to approach the correction of a vibration problem that promises to involve significant structural modifications is through the use of the tools of rational mechanics. A structural math model is to be first be calibrated to simulate the existing vibration characteristics. Modification possibilities are then exercised with the model, and their probability of success is established on paper. In this way the probability of a "one-shot" success when shipboard modifications are subsequently implemented is maximised. The alternative and unenlightened "cut-and-try" approach to the solution of serious ship vibration problems is fraught with frustration, and with the possibility of expending vast amounts of time and money without achieving complete success.

Of course, the paper-studies proposed as a tool for use in correcting a serious ship vibration problem must be concluded quickly. Several months, or even several weeks, is not available when delivery of a vessel is stalled, awaiting the resolution of vibration deficiencies. This places a premium on formulation of the simplest possible structural models which still retain adequate realism to provide the basis for the required judgments as to the relative effects of vessel modifications. This is where the collection of thorough trial vibration data can pay for itself. Measurement of vibratory mode shape data is often a near necessity for securing guidance in formulating calibration models of the desired simplicity, but with sufficient accuracy.

This is illustrated by the following case study.

8.5   Case Study

Excessive vibration of a Type A deckhouse, Chap. 4, Fig. 13, occurs on the builder's trials of an 80,000 DWT product carrier. Vibratory displacement amplitude data are recorded with a multi-channel recorder via phase-calibrated accelerometers mounted at points on the house and on the main deck. The records establish the following information.

(1) The vibration occurs at blade-rate frequency, confirming the propeller as its exciting source,
(2) The vibration amplitude peaks at 100 rpm, and the propeller has five blades. A resonance of the house at 500 cycles per min (cpm) is therefore established,

(3) Vibration records recorded at 100 rpm show that the vibration of the house is fore-and-aft, with fore-and-aft amplitude increasing with a quasi-linear characteristic from low levels at main deck to a maximum of 39 mm/sec at the housetop. The housetop is 15 m above main deck, and

(4) The 100-rpm record also shows that the amplitude of the vertical vibration at main deck is uniform at $V = \omega X = 5$ mm/sec over the house length. The vertical vibration amplitude is also approximately constant at this same level up the house front, which is a continuation of the forward engine room bulkhead.

The above measured characteristics of the vibration under consideration are judged to allow the use of the simple rocking/bending house model in conjunction with the Hirowatari method, Chap. 4, Fig. 13 and Chap. 4, Fig. 14 and Chap. 4, Table 3.

8.5.1 Determination of Model Constants

For a Type A house with $h = 15$ m, the fixed-base fundamental house natural frequency is estimated from Chap. 4, Fig. 14 as $f_\infty = 800$ cpm. Using the Dunkerley formula in Chap. 4, with the measured house natural frequency $f_e = 500$ cpm, the effective rocking frequency is predicted to be:

$$f_R = \sqrt{\frac{1}{\frac{1}{f_e^2} - \frac{1}{f_\infty^2}}} = 640 \text{ cpm}$$

The two frequencies f_∞ and f_R can be used to determine the effective stiffness of the house and its underdeck supporting structure for use in the simple model of Fig. 8.2. For a house mass established as $m = 300\ t$, with a radius of gyration, r, about the house forward lower edge of 10 m, the effective torsional stiffness of the under-deck supporting structure is, from Chap. 4, Sect. 8.4:

$$K_f = (2\pi/60)^2\ f_R^2\ J = 1.35 \times 10^{11}\ N-m/rad$$

where

$$J = m\bar{r}^2 = 3 \times 10^7\ kg-m^2$$

An approximate effective bending/shear stiffness of the house, K_h, is obtained by first lumping the house mass at the radius of gyration above the assumed pin support on the main deck at the forward bulkhead. This preserves the mass moment of inertia in the Fig. 8.2 model. Then, for the house base fixed:

$$K_h = (2\pi/60)^2\ f_\infty^2 m = 2.1 \times 10^9\ N/m$$

Fig. 8.2 Mass-elastic model
of deckhouse and support
structure

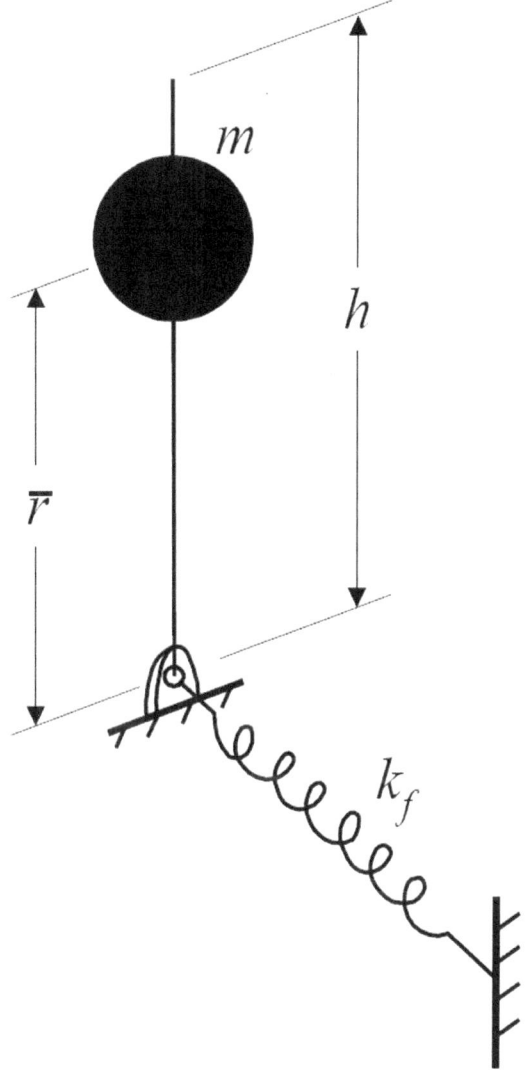

The effective combined torsional stiffness for use in the equivalent reduced one-mass system is then:

$$K = \frac{1}{\frac{1}{K_f} + \frac{1}{(K_h^{-2}r)}} = 0.822 \times 10^{11}\ N - m/rad \qquad (8.1)$$

This combined stiffness can also be deduced directly from the measured natural frequency and the house mass moment of inertia as:

$$K = \left(\frac{2\pi}{60^2}\right) f_e^2 \ J = 0.822 \times 10^{11} \ N - m/rad$$

The effective exciting moment due to the vertical hull girder vibration can be estimated using the following formula:

$$M_e = \omega m \bar{\xi} V \tag{8.2}$$

Here, ωV is the vertical acceleration of the house via the hull girder, so that $m\omega V$ is a vertically downward inertia force acting at the house CG. $\bar{\xi}$ is the longitudinal coordinate to the house CG, measured aft from the house front, which is 5 m. V is the velocity amplitude of the main deck vertical vibration, which is 5 mm/sec. In terms of arbitrary hull girder vibration frequency ω, the exciting moment is estimated as:

$$M_e = 150\omega^2 \ N - m$$

As a final remaining element of the Fig. 8.2 equivalent one-mass model, the damping factor ζ, is estimated as follows using the measured 39 mm/sec house top vibration amplitude. With Φ being the amplitude of the equivalent vibratory rocking angular velocity of the house, the fore-and-aft velocity amplitude of the housetop is approximated as,

$$U = \Phi h$$

where h is the 15 m house height above main deck. Substituting the response formula for the Fig. 8.2 model,

$$U = \frac{\omega h \ M_e/K}{\sqrt{\left[1 - \left(\frac{\omega}{\omega_n}\right)^2\right] + (2\zeta \ \omega/\omega_n)^2}} \tag{8.3}$$

But at resonance, $\omega = \omega_n$, so that:

$$U = \frac{\omega h M_e}{2\zeta} = 39 \ mm/sec \tag{8.4}$$

or

$$\zeta = \frac{\omega h M_e}{2KU}$$

For $\omega = \omega_n = \frac{2\pi f_e}{60} = 52.4 \ rad/sec$:

$$M_e = 4.12 \times 10^5 \ N - m$$

The damping factor is then:

$$\zeta = \frac{52.4 \times 15(4.12 \times 10^5)}{2(0.822 \times 10^{11})(39 \times 10^{-3})} = 0.05$$

With the calibrated model so established as an equivalent one degree of freedom system, with constants, J, K, ζ and M, the above formula can be reused to evaluate changes in the house-top vibratory velocity amplitude, U, resulting from selected changes in the array of design variables included in the simple formulation.

8.5.2   Structural Rectification Analysis

To demonstrate the procedure, stiffening in the form of the added parallel pillars (Chap. 4, Fig. 15) example is contemplated. Following that example, the torsional stiffness of the under-deck supporting structure is raised from the above value of 1.35×10^{11} N-m/rad to 1.475×10^{11} N-m/rad by the pillar addition. From Eq. 8.1, the increased combined stiffness of $K = 0.866 \times 10^{11}$ N-m/rad results in a 2.6 percent increase in natural frequency from the measured value of 500 cpm to 513 cpm. The full power rpm of the vessel is 105, which corresponds to a full power blade-rate exciting frequency of 525 cpm.

The critical has therefore been raised only to a higher level in the operating range from 100 to 102.6 rpm.

At 102.6 rpm, the 5 mm/sec vertical hull girder vibration measured at 100 rpm would be increased by at least the frequency increase cubed, since the propeller exciting force would vary as rpm squared. This is assuming a flat frequency response characteristic of the hull girder (not close to a hull girder critical) as well as a non-cavitating propeller. Assuming a frequency cubed increase in V, the hull girder vertical vibration velocity amplitude becomes:

$$V = 5\left(\frac{102.6}{100^3}\right) = 5.4 \ mm/sec$$

which would still be quite acceptable. However, the exciting moment from Eq. 8.2 would be increased to:

$$M_e = 4.56 \times 10^5 \ N - m$$

at the new resonant frequency $\omega = \omega_n(2\pi)\left(\frac{513}{60}\right) = 53.72$ rad/sec. The house top fore-and-aft vibratory velocity amplitude resulting from the foundation stiffening, assuming the damping factor unchanged, is predicted to be increased from 39 mm/sec to (Fig. 8.3):

$$U = \frac{53.72 \times \frac{15(4.45 \times 10^5)}{0.866} \times 10^{11}}{2(0.05)} = 40 \ mm/sec$$

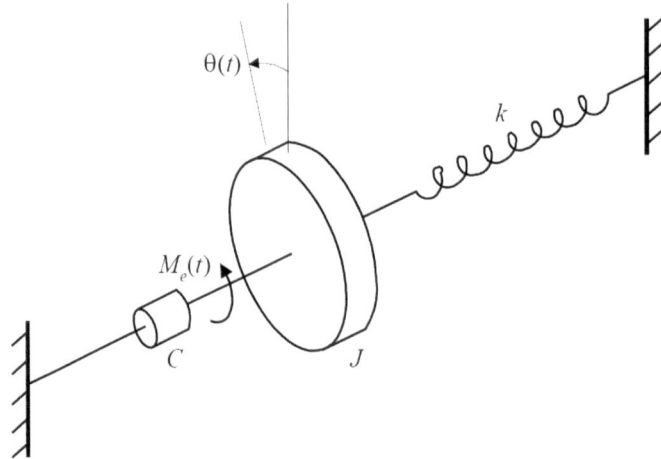

Fig. 8.3 Equivalent one-mass system

This is an increase in vibration of 2.5% over the original 39 mm/sec level. The inadequate stiffening has simply raised the critical to a higher point in the operating range where the excitation is more intense.

Some care is required here in order to achieve a satisfactory result.

In view of the above result, it would be intelligent at this point to evaluate the stiffness increase required in order to achieve a satisfactory vibration level. It is most desirable to move the critical above the full power rpm of 105, rather than decreasing stiffness to get below it. This establishes the exciting frequency at the full power rpm:

$$f = 525 \text{ cpm} = 8.75 \ Hz$$

$$\omega = 8.75(2\pi) = 55 \ rad/sec$$

For crew accommodation areas, the limiting housetop fore-and-aft vibratory velocity amplitude of 5 mm per sec is selected at this frequency. Continuing to assume a frequency-cubed variation in the hull vertical velocity amplitude, the exciting moment is now from Eq. 8.2:

$$M_e = \frac{(55)(3 \times 10^5)(5)(5)\left(\frac{525}{500}\right)^3}{1000} = 4.775 \times 10^5 \ N-m$$

From Eq. 8.3, for $\frac{2\zeta\omega}{\omega_n} << \left[1 - \left(\frac{\omega}{\omega_n}\right)^2\right]$ for $\frac{\omega}{\omega_n} < 1$:

$$U = \frac{\frac{\omega h M_e}{K}}{1 - (\omega/\omega_n)^2} \qquad (8.5)$$

Then with $\omega_n = \sqrt{\frac{K}{J}}$, the required combined stiffness is:

$$K = \frac{\omega h M_e}{U + J\omega^2}$$

Substituting the values, the required combined stiffness is:

$$K = 1.7 \times 10^{11} \ N - m/rad$$

This requires more than doubling the as-built combined effective stiffness of 0.822×10^{11} N-m/rad. Little can normally be done to change the house stiffness; functional requirements of the house usually will not permit the modifications necessary to accomplish any significant increases in house casing section moment of inertia and shear area. Stiffening of the under-deck supporting structure is the only and possibly effective structural modification that can be accommodated. The required K_f, is, from Eq. 8.1:

$$K_f = \frac{1}{\frac{1}{K} - \frac{1}{K_h}\bar{r}^2} = 1.02 \times 10^{12}$$

Therefore, meeting the vibration limit of 5 mm/sec at the housetop will require increasing the effective torsional stiffness of the under-deck supporting structure by a factor of:

$$\frac{1.02 \times 10^{12}}{1.35 \times 10^{11}} = 7.5$$

and this would be impossible to accomplish. For example, if the two parallel pillars of the Chap. 4, Sect. 8.3, example were doubled in number from 2 to 4 and moved 3 m aft to line up under the house after bulkhead, rather than under the house sides (Fig. 6, Chap. 13), K_f would be increased to only:

$$K_f = 1.35 \times 10^{11} + 2(5 \times 10^8)(8)^2$$

$$= 1.99 \times 10^{11} N - m/rad$$

which is still a factor of more than 5 below the required value. At this point, the virtual impossibility of rectifying the problem through structural modifications is clear, and attention would be turned to the propeller.

8.5.3 Propeller Change

Considering an alternative 4-bladed propeller, the critical would be shifted to:

$$100\frac{5}{4} = 125 \ rpm$$

which is 25% above the critical, which meets the rule of thumb. With the foundation unchanged, the housetop vibration at the full power rpm of 105 would be, from Eq. 8.5:

$$U = \frac{\frac{(55)15\left(4.775\times10^5\right)}{0.822\times10^{11}}}{1 - \left(\frac{105}{125}\right)^2} \times 1{,}000 = 16 \ mm/sec$$

The vibration at the housetop is down by almost 50%, but still too high, on the basis of the habitability criterion. Another possibility for the propeller would be to change to 6 blades and lower the critical below full power to:

$$100\left(\frac{5}{6}\right) = 83 \ rpm$$

At full power in this case:

$$U = \frac{\frac{(55)15\left(4.775\times10^5\right)}{0.822\times10^{11}}}{1 - \left(\frac{105}{83}\right)^2} \times 1{,}000 = 8 \ mm/sec$$

This level might be judged to be marginally acceptable as a workspace on the bridge deck at the top of the deckhouse. The disadvantage to 6 blades is the resonance at 83 rpm. At 83 rpm (43.5 rad/sec), the exciting moment, Eq. 8.2, goes down by at least frequency cubed, as has been previously assumed.

$$M_e = 4.775 \times 10^5 \left(\frac{83}{105}\right)^3 = 2.36 \times 10^5 \ N - m$$

With 6 blades, $f_e = \frac{6(83)}{60} = 8.3Hz(52.2rad/sec)$. The resonant amplitude is from Eq. 8.4,

$$U = \frac{\frac{(52.2)15\left(2.36\times10^5\right)}{0.822\times10^{11}}}{2(0.05)} \times 1{,}000 = 22.5 \ mm/sec$$

While this predicted level is excessive, it would not necessarily disqualify a 6-bladed propeller, as continuous operation at any particular lower rpm is usually not critical, and 83 rpm could be simply avoided except in passing.

In this case study, it was decided to change to a 6-bladed propeller. A reconditioned propeller of the correct diameter and pitch was refitted on the vessel. It was confirmed in advance that the blade area met the Burrill Criteria 10% back cavitation as suggested at Chap. 4, Fig. 14. The deckhouse vibration was judged by the owner to be acceptable in closely enough meeting the habitability criteria from Chap. 7, Table 9. A vibration increases in passing through the 80–85 rpm range was detectable by personnel in the deckhouse, but it was quite moderate and judged also to be acceptable. Successful acceptance trials were completed within the delivery guarantee margin mandated by the contract.

9.1   Introduction

Ship flexural vibration excited by the seaway can be any of three types:

(1) Springing,
(2) Bow flare slamming, and
(3) Bottom slamming.

9.2   Springing

Springing is a continuous-type vibration of the hull excited by the short wave (high frequency) wave components of the seaway. This is versus transient slamming, which is
addressed in the next section. The most well-known cases of springing occurred on the
Great Lakes Ore Carriers built in the 1970's. There, the locks limitations on vessel beam
and depth required longer and longer ship lengths to meet the desired cargo carrying
capacity. Subsequently, the ships had a very low 2-noded vertical flexural natural frequency. This, coupled with the short-length (high frequency) spectrum of the Great Lakes
seaway, resulted in measurable response magnification in the first flexural mode. The level
of the flexural response could, however, not be explained by sub-resonant magnification
alone. It was concluded through investigations by ABS that nonlinear hydrodynamics in
producing low frequency "difference wave" excitation was partly responsible and therefore played a role. Although carefully monitored for some years, no failures of the Great
Lakes bulk carriers have ever been attributed to springing. Springing could however be
brought back to the forefront of attention by increasing vessel speed. The higher the speed
the higher the "frequency of encounter" with the sea waves. The result could be the same

© The Author(s), under exclusive license to Springer Nature Switzerland AG 2025   101
F. Karkori, *Ship Vibration 1*, Synthesis Lectures on Ocean Systems Engineering,
https://doi.org/10.1007/978-3-031-75072-4_9

Fig. 9.1 Great lakes ore carrier 'EDMUND FITZGERALD'

as appeared on the Great Lakes some 30 to 40 years ago, but for a different reason. Care needs to be exercised in designing the new generation of high-speed ships with enough hull girder stiffness to avoid significant dynamic magnification of hull flexure by high frequency of encounter relative to sea waves (Fig. 9.1).

9.3   Bow Flare Slamming and Whipping

Bow flare slamming occurs due to flattening of the bow (and stern) lines toward horizontal in the vessel body plan. Here the bow flare slamming tends to be more continuous and of longer duration than fore-foot bottom impact slamming. Therefore, for large container carriers as a typical example, the bow flare slamming can produce significant transient hull girder vibratory response called 'whipping', which usually cause the amplification of global hull girder bending loads. The bow flare slamming is also known to be capable of producing local damage to bulwarks and side plating in the flare regions. The inclination is to add flare in the bow to provide the hydrodynamic stiffness to avoid bow submergence in the waves. But the price paid in bow deceleration to prevent submergence is high flare loads and potential structural damage, both globally and locally. Usual practice is that with high-speed ships, flare angles of the bow side shell and bulwarks are not to exceed angles of 40 to 45 degrees relative to the vertical.

9.4   Bottom Impact Slamming

Emergence of the fore-foot bottom above the water surface followed by re-entry tends to produce very intense slamming that occurs for a short duration (as little as fractions of a second). Here the essentially instantaneous application of basically a point force will excite the ship to "ring" in all of its normal modes simultaneously, but to different degrees of course, depending on the spectral content of the slam impulse. This can be particularly serious with very low natural frequency substructures base excited through the transients from the hull girder. It can also produce local plating damage in the fore-foot due to the high hydrodynamic impact pressure loadings. Aside from designing for a deep forefoot, the best protection against bottom slamming damage is the adjustment of seaway operations to suit the local sea characteristics.

9.5   Concept Design Checklist

The following provides a summary of the check list that is to be encountered in the typical concept design process.

(1) During the initial engine selection, check the second order vertical moment M_{2v} from the potential vendor, by the procedure outlined in Chap. 3. If the PRU (power related unbalance) exceeds 220 N-m/kW, consider either change of engine selection or installation of moment compensators. Also, the installation of engine lateral stays on the engine room structure is to be addressed at the early design stage,

(2) During the initial stern lines fairing and propulsion system arrangement studies, check the maximum angles and minimum propeller clearances in Chap. 3, Fig. 5 or Chap. 3, Fig. 6,

(3) Check if the aperture clearance provided above is adequate to accommodate a propeller diameter satisfying the Burrill cavitation criteria for a maximum of 10% back cavitation, as indicated in Chap. 3, Fig. 8.

If either of the points at (2) or (3) are not satisfied, plans are to be made to perform detailed wake and propeller cavitation studies based on model testing and/or numerical simulation. Once preliminary hull principal dimensions and weight are established, check the vertical hull girder natural frequencies up to at least five nodes to avoid the resonance or near resonance (20% separation minimum) with the engine second-order vertical moment at full power. This may be achieved through the procedure outlined in Chap. 4, Sect. 9.3. If twice the rpm of the engine at full power falls within 20% of any of the estimated natural frequencies, the more precise FE-based free vibration analysis of the hull girder natural frequencies is recommended. At the time of propulsion machinery and foundation arrangement design, follow the procedure outlined in Chap. 4, Sect. 9.5

to check the thrust bearing structural foundation stiffness needed to avoid blade-rate resonance with either of the first two system longitudinal natural modes in the upper power range. If this cannot be achieved, the change of propeller blade number and/or rpm is to be considered so that a minimum 20% margin on the full power P_{rpm} is provided. In developing the scantlings and supporting structure of the deckhouse, perform the analysis of Chap. 4, Sect. 7 to avoid fore-and-aft blade-rate resonance, within the 20% margin at full power.

Correction to:
F. Karkori, *Ship Vibration 1*, Synthesis Lectures on Ocean Systems Engineering,
https://doi.org/10.1007/978-3-031-75072-4

This book contains overlap in text with the previously published content [1] that was inadvertently omitted. The authors failed to attribute the reference [1]. The authors have now obtained permission to re-use this content from the American Bureau of Shipping.

Where [1] is: American Bureau of Shipping (2024), Rules and Guides https://ww2.eagle.org/en/rules-and-resources/rules-and-guides.html

---

The updated version of this book can be found at
https://doi.org/10.1007/978-3-031-75072-4

F. Karkori, *Ship Vibration 1*, Synthesis Lectures on Ocean Systems Engineering,
https://doi.org/10.1007/978-3-031-75072-4_10

References

General

SSC-350 (1990) Ship Vibration Design Guide

Concept Design

Burrill LC (1934–35) Ship Vibration; Simple Methods of Estimating Critical Frequencies," Transactions, NECI, Vol 51

Burrill LC (1943) Developments in Propeller design and manufacture for merchant ships. Transactions, IME London, Vol 55

Harrington RL (1992) Marine Engineering, SNAME

Hirowatari T, Matsumoto K (1969) On the Fore-and-Aft Vibration of Superstructure Located at Aftship (Second Report). JSNA Trans 125, June

Holden KO, Fagerjord O, Frostad R (1980) Early design-stage approach to reducing hull surface force due to propeller cavitation. SNAME Trans

Hylarides S (1978) Some hydrodynamic considerations of propeller-induced ship vibrations. Ship Vibration Symposium, 78, Washington, D.C., October

Johannessen H, Skaar KT (1980) Guidelines for prevention of excessive ship vibration. SNAME Trans 88

Kumai T (1968) On the estimation of natural frequencies of vertical vibration of ships. Report Res Inst Appl Mech 16(54)

Sandstrom RE, Smith NP (1979) Eigenvalue analysis as an approach to the prediction of global vibration of deckhouse structures. SNAME Hampton Roads Section Meeting, October

Schlottmann G, Winkelmann J, Sideris D (1999) Vibrations of resilient mounted engines Jahrbuch der Schiffbautechnischen Gesellschaft 93:185-193 (In German)

SNAME (1988) Principles of naval architecture, Vol II, Chap 7 "Vibration"

S Bros 1977 RND marine diesel engines, technical data Sulzer Bros. Ltd. Winterthur, Switzerland, October

WT Thomson 1973 Theory of vibration with applications Prentice-Hall

Veritec (1985) Vibration control in ships.

© The Editor(s) (if applicable) and The Author(s), under exclusive license to Springer Nature Switzerland AG 2025
F. Karkori, *Ship Vibration 1*, Synthesis Lectures on Ocean Systems Engineering, https://doi.org/10.1007/978-3-031-75072-4

FE Analysis

American Bureau of Shipping (2005) Ship Vibration Analysis Procedure Guide
American Bureau of Shipping (2001) Report RD 2001–11 Program Development for Added Mass
    and Buoyancy Spring
American Bureau of Shipping (2005) Report RPD2005–07 an integrated computational process for
    Cavitating propeller induced loads
SSC-387 (1996) Guideline for evaluation of finite elements and results

Measurement

ANSI S2.27 (2002) Guidelines for the measurement and evaluation of vibration of ship propulsion
    machinery. American National Standard
BS 6841 (1987) Guide to measurement and evaluation of human exposure to whole-body mechanical
    vibration and repeated shock
ISO 10816 (2001) Mechanical vibration – Evaluation of machinery vibration by measurements on
    nonrotating parts
ISO 2631–1 (1997) Evaluation of human exposure to whole-body vibration - Part 1: General require-
    ments
ISO 2631–2 (1997) Evaluation of human exposure to whole-body vibration - Part 2: Continuous and
    shock induced vibration in buildings (1 to 80 Hz)
ISO 4867 (1984) Code for the measurement and reporting of shipboard vibration data
ISO 4868 (1984) Code for the measurement and reporting of local vibration data of ship structures
    and equipment
ISO 6954 (1984) Mechanical vibration and shock- Guidelines for the overall evaluation of vibration
    in merchant ships
ISO 6954 (2000) Mechanical vibration - Guidelines for the measurement, reporting and evaluation
    of vibration with regard to habitability on passenger and merchant ships
ISO 7919 (2005) Mechanical vibration – Evaluation of machine vibration by measurements on
    rotating shafts
SNAME Technical and Research Bulletin No. 2–29 (1993) Guide for the analysis and evaluation of
    shipboard hull vibration data, Jersey City, NJ
SNAME Technical and Research Bulletin No. 2–29A (2004) Measurement and evaluation of struc-
    tural & machinery vibration in ships, Jersey City, NJ